GOLD NUGGETS

Readings for Experiential Education

Edited by

JIM SCHOEL

and

MIKE STRATTON

Project Adventure, Inc. Hamilton, MA,
Covington, GA and Portland, OR
© 1990 by Project Adventure, Inc. All rights reserved
Printed in the United States of America

95 94 93 92 91   7 6 5 4 3 2 1

ISBN 0-934387-09-5

Complete copyright information and acknowledgements are
located on pages 155 through 160.

This book is dedicated to its contributors, those teachers and leaders who have experienced the power of a well-placed reading.

CONTENTS

SPECIAL APPRECIATION

When we started work on this book, we didn't know for certain that we'd need to get permission to use material that was not public domain, although we had some suspicions about it. Tom Zierk confirmed those suspicions, which meant that a lot of research had to be done, and letters written. Sue Schaefer, through her use of readings at the Gordon College La Vida program, became interested in the project and applied to PA for an internship. She didn't know that she would be spending many hours in the Boston Public Library trying to find the book and page number of some obscure writer's work. The research and permission work is almost entirely of her doing, along with a whole range of contributions and comments from her own files. What began as a 5-month commitment grew into one and one-half years. Without her work we would not have been able to publish this book.

PREFACE

Why are readings important? Short messages that grab the attention, the imagination, the heart and spirit, are helpful teaching tools, and many leaders use them to frame and reinforce what they are trying to convey through other mediums. They are part of a profound oral tradition that has its roots in sermons, political discourse, debate and storytelling. For teachers and Adventure leaders they can provide a crucial shift to another level of thinking where, with perhaps one line or a simple story, the unreasonable becomes reasonable, and a group or an individual is able to gain insight into the activity at hand. Activity-based programs provide us with many teachable moments, when we are provided opportunities to offer a thoughtful insight relating to the overall setting — location, activity, subject matter, group situation, etc... Our shortened attention span brought on by the instant gratification/TV generation makes us scramble for these precious moments. A limited attention span is a good reason to use readings. The messages they convey can work at shutter speed to open up a wide range of thinking. Certainly, longer essays and books need to be taught, but one sentence, or a paragraph, or a well-selected story can have the impact of longer material. Too often we measure content by volume. Besides, when readings are connected to a physical experience, the volume is the reality of what is taking place, with the short reading being a page or chapter of the whole experiential tome. Readings require scrutiny, because, over time, people have found a continual well of meaning coming from them. In this sense readings are jewels — that mountain stream flowing gently into a meadow in the Olympic mountains…perhaps nothing huge and dramatic, but the doe and her two fawns over there somehow symbolize something. That's what a reading does. It frames our experience, giving us life symbols to help us deal with our present reality.

We can go further by studying readings in a class, breaking them down, looking deep within them. Study can illustrate the connection between words and what the words represent, between the thought and the act. When we provide our students with those bridges, that's good teaching!

I've used readings in my own teaching for many years. I've read Gary Snyder on a kitchen midden on a Maine island, Robert Frost on a snowy birch-lined trail in New Hampshire, Chief Seattle in a classroom in Gloucester, and Rafe Parker during a Project Adventure Counseling workshop. As I look back, the readings always had an impact on someone, if not everyone.

Given my own high level of disorganization, I found that though I was able to keep track of readings that were supremely important to me, I had difficulty collecting a broad range of them and keeping them in one place. Readings on the run would get wet, lost, dirty, or fail to make the trip altogether, leading to great frustration. A gathering place became a necessity, if only for my own purposes.

It was while beginning work on the book *Islands of Healing* that I decided to put together a book of readings. My personality leads me into deep trouble in terms of commitments, and this was another of those megalomanic launches into infinity. "If one book, then why not two?" Of course we hadn't finished the first book, so I was yet to experience multiple deadlines, rewrites and days of glazed-over eyes and melted brain.

During this time I reconnected with a friend I'd met at Hurricane Island Outward Bound School (HIOBS), Mike Stratton, or "Strats." At the invitation of Lance Lee, one of Mike's closest friends, I became involved in Mike's cancer support group. He was diagnosed as having a brain tumor that summer and was very depressed. One of the purposes of the group was to find ways to help Mike get "off the couch." Mike was a true pioneer in the field of Experiential Education, first as an Outward Bound student (his daisy eating scene while on solo, documented in the Colorado Outward Bound School movie, is a classic of experiential lore), then as an O.B. instructor. He later developed the Bounders program at the Carroll School in Lincoln, Mass., a school primarily working with young dyslexic students.

He was a staunch supporter of the Association for Experiential Education, receiving its Practitioner of the Year award in 1986. He was a public voice through television appearances, his book, *The Bounder's Bible*, and by his constant availability to anyone wanting to learn how to do Boundering. A quote by one of his students underlines some of the values he espoused as a teacher: "What was the hardest part of the Bounders program? Sleeping in a tree! If you could do the course again, what would you do differently? Sleep in the tree longer!" Finally, he was an extraordinary individual, connecting personally with those he met, sharing, listening, caring, being a pal, bringing presents, being thoughtful, cracking a joke, reminiscing about the wonderful and unique people we've known.

I remember the first of Mike's cancer support group meetings I attended not so much for the content as for the intense headache I left with. Here was a young man with a young family. He was at the peak of his profession with people all over loving him and looking to him for leadership and inspiration. And he had a brain tumor. Since my mother had died just the year before of another form of cancer, the conflict was heightened. "What can I do?" I mulled as I drove home in the rain.

An idea came to me the next day. Why not approach Mike with a mutual project, one that would benefit both of us, and give him something specific to work on? He'd been depressed about everything, hadn't been able to work or even drive a car. I broached the idea with him, he liked it, and the concept of co-authorship of a book of readings was born. And for me, two years of a close relationship with Mike Stratton began. We named the book *Gold Nuggets*, because Mike had roots in Golden, Colorado (the title also relates to another Project Adventure publication, *Silver Bullets*, by Karl Rohnke). We went through our files — cardboard boxes of scraps. We called friends in the field of experiential education for their favorite readings, asking them to frame their submissions with personal insights. We did a Project Adventure Advanced Adventure Based Counseling workshop together, rowing around Cape Ann in the process. We got dizzy with intensity for nuggets we came across, and for the feelings developing between us.

I can reach back into all of that now. Today, as I write this, I came across a reading Mike gave at a funeral for a student we both taught. (Tommy Sargent, who'd set the record for number of days spent in the Carroll School bomb shelter, a notable Mike Stratton Bounder's solo experience, had been killed in a Gloucester motorcycle accident.)

> Do not stand by my grave and weep.
> I am not there, I do not sleep.
> I am a thousand winds that blow.
> I am the diamond glints on the snow.
> I am the sunlight on ripened grain
> I am the gentle autumn rain.
> When you wake in the morning hush
> I am the swift uplifting rush
> of quiet bird encircled flight.
> I am the soft star that shines at night.
> Do not stand at my grave and cry,
> I am not there. I did not die!

I realize now, looking at this scrap of paper from his file, that Mike meant this poem for himself as well. When I last saw Mike in the fall of 1987, right before he died, he said to me, "Keep the work going, Jimmy. Keep it going!"

It is three years later. Finally, with crucial and timely help from Sue Schaefer (editing, entering, research, permission letters, and inspired commitment), *Gold Nuggets* is underway again.

Jim Schoel
September, 1990

Mike Stratton

At the 14th National AEE conference (Association for Experiential Education), Mike Stratton presented a workshop on the use of readings. This was a special conference for Mike, for it was here that he received the AEE Practitioner of the Year award. The following is an excerpt from a tape of his workshop:

"This is probably the best reading of all (he holds up a blank page). Those of you who have nothing probably have the best reading of all, because often a reading can spoil or interrupt what is all around you that is more beautiful than any reading. So 'travel light, freeze at night.' We work with people, not paper. We don't want to overload ourselves with paper.

"There are several ways you can go: a blank journal, a coffee-table book, or your own collection of newspaper and magazine articles. The Boston Globe or any newspaper is a great source. I happened to live with a guy named Lance Lee, on a sailing ship, a square-rigged ship, and every morning we had (given by Lance) readings. Some of us know a crazy guy from Hurricane Island named Peter Coburn, who contributed a lot of nice readings I want to share with you. He has a book he's had since about 1966, he has it in a leather-bound case — carries it everywhere. He carried it a couple of weeks ago for a one-night trip, carried the whole thing. The book has capsized at sea, it's been rescued. I asked Peter for some of his readings, and his response was, 'Hey, I spent a long time making my own. Categorizing, documenting, rewriting, and that was half the fun.' It means a lot more (to do it yourself), so to hand students one of these (readings books) on day one, or even at the end of a course, I would not feel that that would be appropriate. I would have students make their own, do some drawings too. All of these readings go back to the beginnings of communication. There weren't Xerox machines and televisions, and everything. They transmitted things by symbols where they lived and eventually tablets of some sort, but mainly it was an oral tradition. I'm sure you'd agree that we all enjoy being read to or told a story. The aspect of being read to goes way back — it's in us. There is definitely an appeal for most people to listen and respond and think about these things. It's maybe harder today than it was years ago. That's why making your own reading book is a challenge. Anyone can fork out money for this coffee table one, and I urge you to do so, but it's much better to make your own and have your students make their own. When I was working with dyslexic students, I had them copy readings. I'd give a reading to them, but I'd have them copy it at home, so they had to

practice the penmanship, the spelling, and we had a lot of discussion about them.

"A symbolic way to show the use of readings would be the extension cord. Say the experience is over in another state, and now I have a chance to think about it, or the group does. We need extension cords to keep things, the juice, flowing. Readings serve that purpose. They are extenders back to the memory of the experience, and they can be the inspiring juice that's going out prior to an experience. To me, that's going the extra yard as a teacher or an instructor, to make up (duplicate) one of these readings to send home with them, because these become that extension cord. They have something tangible. They might have a badge, they might have a certificate, they might have a T-shirt, but this might be more appropriate than any of those, because they heard it at an appropriate time or they copied it themselves.

"It all starts with you having a personal belief in these things. You use them on yourself. You spit them out when you're running or climbing. Readings come from the heart and the head. And a combination of everything: head, heart and hands, if you're a hands-on type."

At the closure of the workshop, everyone went outside and held hands. Mike summarized: "These readings are meant to reach out and touch, and once the students go home this may be the only way you can touch them for the rest of their lives."

Mike Stratton
October, 1986

INTRODUCTION

Nugget Selection

Dropping a nugget into a group can be fun, dramatic and effective. But there are times when I've done a reading and seen the quizzical look on the faces of the kids in front of me — "Where is he coming from? Just who was that masked man?" and, "How does that fit?" I'm getting at the sometimes random, sometimes static use of nuggets I've fallen into. It's not that I didn't think about how the nugget applied to a certain situation. The Frost poem *The Road Not Taken* was carefully selected to coincide with a winter trail junction where there was a real choice, one that would hopefully stimulate the students to think about their own life choices. But there may have been times, when I look back, that another reading would have worked better for what that particular group was experiencing at that particular time. *The Road Not Taken* was good, but something else may have been even better. Because I had done it before, it seemed right to keep on doing it, without question. We all know into what kind of dead end that type of teaching leads us.

So I've begun to think a little bit more about the nuggets I use and when I use them. This thinking has been stimulated by the activities selection work we have done with Adventure Based Counseling. Coming up with a group responsive formula has forced us to be clear about what activities we select in relation to the specific issues a group may be dealing with. The concept of sequencing keeps coming up: You do certain things before you do others. Yes, you can plan and lay out an extensive scenario, but you also need to adjust to what is going on at any given time. The planning and adjustment needed some strategies that would allow us to be more thoughtful about how we lead our groups.

Out of this desire to be more thoughtful about our Nuggetting, we have divided our Nugget collection into categories: Commitment, Community, Humor, Journey, Leadership and Values. Here's our definition of each category.

Commitment

During the adventure experience we encourage people to set goals and to reach for them. This energetic pursuit of one's choices requires a strong dose of commitment. In fact, it is impossible to achieve one's goals without committing oneself to them. It's a strong word, and unequivocal. I once wrote a poem entitled *To Build a Wooden Boat* (pg. 5). I shared it with a boat-building class that I was teaching at the time. The final line states, "You either do it, or you don't." The kids understood just what I was talking about, for the issues around boat-building are uncompromising. You may fuss about and pretend to build a boat, but in the end it either floats, looks like something, and is worth something, or it is unfinished, or finished poorly, or sinks. Either/or. In my work with kids I've seen a thirst for such a lack of equivocation. "These are the issues. I've decided that this is what I want to do. Now let's go and do it." It may take some hard work to get to those commitment points, and a mistake often made is slapping people with them too soon or overwhelming them with too many choices.

Community

When Shelia Torbert reads her anti-drug poem *Beware of Me, My Friend* (pg. 23) she is penetrating into the essence of community — taking responsibility for each other. We don't need to be oppressive about it, but in the areas where we are interconnected, we need to learn how to make those responsibilities viable. It's not an easy thing to do. Fear, pride or simply laziness get in the way of making an effort to cross the lines and get involved.

Building a sense of responsibility for each other is an

essential component of group work, and it is our assumption that if it is experienced and translated in one place, it is more likely to be practiced in another. Community responsibility is multifaceted. It can be symbolized by the reading, *The Monkey Rope* (pg. 7), it is seen in the rage of Randolph Bayliss in his letter concerning the botched oil clean-up effort in Prince William Sound, Alaska (pg. 34); it is also apparent in M. Scott Peck's assertion that "a community is a group that can fight gracefully." That's why it is important to give the group time to think not only about how the reading connects to the current activity but also how it relates to the world at large.

The group itself is, of course, our best reflection of community, for it is something upon which we can have an impact. Too often we preach at our groups about issues they are powerless to solve. We need to talk about the oil spill. But we also need to look at the issues right in front of us. Then the message of *Enviropimp* (a way the author describes his occupation, which involves drafting Environmental Impact Statements) can really speak to us.

Humor

A quote Bonnie Hannable gave me has been coming across my screen lately — "Life is serious, but not that serious." Now that I've settled in for the long haul, raising a family and keeping as many balls in the air as possible while trying to pay the bills, the tension can be overwhelming at times. Its important to keep a proper perspective. I mean, give me a break! Being too tight can not only injure one's health, it's not any fun either! Each of us was once a kid and kids just love to play and laugh. Unfortunately, we lose that ability all too quickly. I know that with my Judeo-Christian, co-dependent concern for the world it has been difficult at times to let go, so much so that when I do, the muscles in the back of my head hurt because they are being forced to exercise (at least that's my theory).

It's easy to poke (or stab) fun at the expense of others, however. Some of the funniest things I've heard are those types of jokes. But when I really take a look at it, some of those feelings aren't really very satisfying. Much of humor is a power play — one that leads to put-downs and discounting, or "dising." How do you laugh with someone without laughing at them? Being connected to your feelings is one way. Decisions regarding appropriateness come from our own inner sense of things.

I suppose *The Cremation of Sam McGee* (pg. 52), where an Alaskan prospector fulfills his friend's final wish to be warm "just like back home in Tennessee," could offend some people, but it never fails to get a hearty laugh, especially when read to a group that has undergone great mutual effort in the cold. It's always been a good risk. On the other hand, you can go so far in an explanation of Italian Golf (a Project Adventure activity), that the offense can become intolerable. So a fine line has to be walked. If you cross over it and are confronted about it, don't try to cover your tracks. "I tried, I see your point," is a necessary response. Keep on trying and learning.

Journey

We are all on a journey. When I was a kid I read a book entitled *Sailing Alone Around the World*, by Joshua Slocum. This Massachusetts native was the first person to complete such a trip, and this in the last century when there were no radios, much less satellite navigation systems. His voyage has stood as my image of a successful journey,and he kept a journal, writing diligently about what he was doing while confronting all sorts of difficult and dangerous odds. His trip through the Straits of Magellan stands as one of the most exciting pieces of travel literature ever written. (I should say journey literature, and note that the term journal is connected to the term journey.) Ever since I read his book, I have vowed to some day journey through those same straits in a small boat.

But my point is larger than the physical journey where one "meets rare beasts and has unique adventures." I have had to learn that an individual's journey is infinitely more complex than that. The rare beasts can be right in one's backyard, where a man learns how be a good father by giving up his own sense of time and allowing himself to be soaked up in the time-sense of his child. What a relief such an act can be, for on his journey he has allowed the small but viable and rich voice of this little person to penetrate and be part of his path. He is no longer alone on his journey.

We don't have to be heroic like Odysseus to go on a journey. We can be heroes in our own right and in our own time. I read an article about a husband who had spent six years with his wife battling cancer. He said about his wife that, "her trip was cancer," meaning that she found self-fulfillment in the battle in terms of all that it demanded from both of them — emotionally, spiritually, physically. That was the lot she had to deal with, so she took it to the limit.

We all need to look at ourselves more closely as individuals, something that our mass culture tries to take from us because of the constant barrage of information encouraging us to be like everyone else. I used to help write a magazine with students. I remember one fellow we interviewed in Gloucester whose name was Kleimola. He was a Finnish stone cutter and was convinced that he had nothing to say. But as we got into his everyday experiences of cutting stone in the Lanesville quarries, he realized that he had a great deal of information that few, if any, other people had. He was really sold when he saw his article in print! What he experienced is an essential part of the process of valuing, or the expanding circle of value that we should be willing to look for and accept from our experience. His life experience was a journey.

Leadership

Much of the current educational jargon revolves around the need to develop strong leaders — peer leaders, natural leaders, student body leaders, team captains. The need for leaders can cause teachers to pick them out and get them on line without any real training, goals or vision. Leadership in the Adventure world should be an active, goal-oriented position where there are specific activities for students to practice on a regular basis. This counters an inertia and frustration that can develop: "Don't just stand there, lead something!" In developing leadership in young people, it's important to have them feel special, and there is no better way to do this than to give them plenty of time to talk about their experiences. This talk time includes helping the leaders set and monitor goals, generally staying with them as they develop a sense of purpose and community. It is in that active context where readings about leadership are most effective.

How one leads is a part of that training, for there are many different approaches. In Terry Dobson's *Aikido in Action* (pg. 94), a young man is forced to consider the essence of his three years of Aikido training. What he observes runs counter to all his desires. This reading can stimulate people to consider that perhaps there are often more effective ways to get results. Looking for one's own leadership style, rather than trying to mimic someone else's, is a necessity. A leader can learn from others but must ultimately settle on an approach that is uniquely personal.

Values

Being preached at about values is a common complaint, and it would be a temptation to use nuggetting for that purpose. Paul Radcliffe has a simple formula that mocks the preachy attitude: "Be good. Avoid evil." He's said it so many times when we were taking leave from each other, and with such an unpreachy affect, that I always think about it (whether I want to or not).

Project Adventure runs the preaching gauntlet when we ask people to adhere to the Full Value Contract. A major tenet of the contract is that all group participants are valued for who they are, not for who the group may or may not want them to be. Care must be taken in presenting the contract, however, lest it be just another in a long line of lectures about the obvious. If not, we will be left with the same number of listeners that the flight attendant gets for instructions about oxygen masks. Situations demanding clarification are the best teachers. The Full Value Contract has such practical applications in those situations, that its importance becomes increasingly obvious. It is a framework that works. It is not just a bunch of jargon. A reading can reinforce the values that emerge from actual situations, rather than being something superimposed from the outside.

When we hear Henry David Thoreau say, "I went to the woods because I wished to live deliberately, to front only the essential facts of life, and see if I could not learn what it had to teach, and not, when I came to die, discover that I had not lived," (pg. 117) he is talking about going deep into life's experience. He wants a fullness, a depth of understanding, rather than an experience of the mundane. He is willing to do this because of the value he has developed for life. It is a choice that we can make and help the participants in our groups explore. It is not a choice about whether to turn right or left but about the quality of our experience. Do we want the best, or are we willing to settle for the other?

Strats' Tower of Strength (pg. 126), where Mike defines the five C's of Courage, Competence, Culture, Courtesy and Character, sums up in a simple, straightforward manner the significant values that he looked to in his life.

One-Liners and Songs

The final two sections are not organized around specific themes. Nuggets are placed in these sections because of the form they take; i.e., short stabs of meaning, or singing activity. The one-liners have their own categories: *Growth, Outlook, Relationships*, and *Taking Risks*.

Model Nuggets

We want to enrich your Nuggeting with the experience of a veteran group of teachers, outdoor leaders, and counselors who have used readings to enhance their work. These Model Nuggets are placed at the beginning of each section. Contributors first present their reading of choice, then follow it up with a framing discussion, which can cover a range of issues — where it was used, why it was used, what kind of questions they would ask, and what it means to them. I hope their experiences will give you some ideas about how to use the offerings in this book, and to help you develop your own readings repository. In fact, this is a central purpose of *Gold Nuggets* — to get you, the reader, to begin to collect your own readings for your own specific area of interest. Mike called this process BYOB (Build Your Own Book). We have had some contributors balk at giving their readings to a book because they feel it takes away from the power of each unique individual's collection process. While there is some truth in their reluctance, we also know that knowledge is progressive, and that people can learn from our collection. But it must be dynamic! To take this book and only use the contents of it without developing your own model nuggets is to miss the most exciting part of Nuggeting. Discovering a new reading, thinking through where it should be applied, then experimenting with it is very much like inventing a new Adventure activity or a variation of an old one.

Please note that as we present these choice readings, we are aware of the limitations we must work under, namely that readings are part of an oral tradition. Because of this, each time they are used they come across differently, embellished according to the imagination of the reader, and the needs and responses of each group. Capturing that sense is what we want to achieve.

Here is one Strats' Model Nuggets:

The Second Cigar
William Barrett

His own (Kierkegaard's) explanation of his point of departure as a thinker is given in a characteristically vivid and Kierkegaardian passage in the *Concluding Unscientific Postscript.* While he sat one Sunday afternoon in the Fredericksberg Garden in Copenhagen smoking a cigar as was his habit, and turning over a great many things in his mind, he suddenly reflected that he had as yet made no career for himself whereas everywhere around him he saw the men of his age becoming celebrated, establishing themselves as renowned benefactors of mankind. They were benefactors because all their efforts were directed at making life easier for the rest of mankind, whether materially by constructing railroads, steamboats, or telegraph lines, or intellectually by publishing easy compendiums to universal knowledge, or — most audacious of all — spiritually by showing how thought itself could make spiritual existence systematically easier and easier. Kierkegaard's cigar burned down, he lighted another, the train of reflection held him. It occured to him then that since everyone was engaged everywhere in making things easy, perhaps someone might be needed to make things hard again; that life might become so easy that people would want the difficult back again; and that this might be a career and destiny for him.

Irrational Man: A Study in Existential Philosophy

Mike's Comments: This reading about the great philosopher Kierkegaard's choice or need to make a choice has always enchanted me. Was he going through a

*midlife crisis? Or, more than likely, alarmed at the
acceleration of the industrial revolution (high tech) at the
expense of traditional ways and crafts, old values, and a
sane lifestyle.*

*I have used this reading when opinions about an
expedition begin to overshadow the decision making
process. ("This sucks!!! We could do this much faster
with a snowmobile instead of these damn snowshoes."
Also using a motor instead of oars and sail, MacDonalds
vs. a quality outdoor feast, and, "I miss my toilet!")*

*These moments offer great opportunities to discuss
why you are out there doing it vs. home watching Holly-
wood adventures. They are also a good chance to
reconnect with the good old days and ways of life,
history, lifestyles, and struggles).*

*I like to combine this with John Ciardi's quote:
"Every game ever invented by man has strict rules to
make it challenging and fun."*

*The fun and meaningfulness come from making the
hard look easy. Just try to imagine a tennis game without
a net and lines.*

*So who is right? Mickey Mouse who says, "Life is a
contest to see who can do the least!" Or Helen Keller who
said, "Life is an adventure or nothing."*

Another Model Nugget from Mike Stratton is the *Tucker
Foundation Credo*, which was on a wooden sign at the
entrance of College Hall at Dartmouth College. (See pg. 71)

GRABBS

Now that you have the nuggets in categories, the trick is
how to get to them when you need them. This is where
GRABBS comes in. Now I'm willing to admit that acronyms
are boring (unless, of course, you've invented one), but
please bear with me — it just might work for you. At least
this one is active, because it makes you really interact with
your group in order to find out where they are. You've got to
reach out and GRABBS someone, not just stand there and
passively spot, or observe, or walk along with them, or belay
them while thinking about something else. GRABBS is a tool
for reading or assessing a group's needs at any given moment. It was developed as part of an overall activities
selection and adjustment process which is outlined in the
Project Adventure counseling book *Islands of Healing*. It can
be used when laying out a group's curriculum, and it can be
applied on the run when making necessary adjustments to
that curriculum. Since all classes and groups vary consider-
ably, GRABBS is an attempt to do battle with static responses from leaders. Even experiential educators can get

locked into doing things one way! So we have found it useful to apply GRABBS to our nugget selection as well as to our activity selection.

Here's the breakdown of each part of GRABBS:

Goals — How does the activity relate to the group and individual goals that have been set or that you see emerging?

Readiness — This regards levels of instruction and safety. Is the group ready to do the activity? Will individuals endanger themselves and others? Do they have the ability to attempt or complete the activity? What will you have to do to change the event to compensate for lack of readiness?

Affect — What is the feeling of the group? What kinds of sensations are they having? What is the level of empathy or caring within the group?

Behavior — How is the group acting? Are they resistive? Disruptive? Agreeable? Are they more self-involved, or group-involved? Are there interactions affecting the group, both positive and negative? How cooperative are they?

Body — What kind of physical shape are they in? How tired are they? Do they substance abuse? Are they on medication? How do they see their own bodies?

Stage — At which developmental stage is the group? Groups will go through many levels of functioning, and no level is static. Having a schema to describe these levels will provide you with another means of assessment (Forming, Storming, Norming, Transforming). Recycling between Storming and Norming is a continuous process and an essential part of the growth.

The effort for us as we go a nuggetting is to use GRABBS in reading our group. Out of this, we can choose to zero in on some individual issues, or we can focus on the group as a whole. For example, when dealing with **Goals**, if the essential goal is to trust others, you can go to Community for *Friendship* (pg. 35) by Kahlil Gibran. The statement that "your friend is your needs answered" can show the importance of speaking your mind, for true friendship is honest. Only through honest give and take can true growth occur, but it must take place in an arena of trust. The reading *Listen With Your Ears* (pg. 38) looks at trust from the angle of a frightened, isolated person who is asking for more than educated advice. If a goal on the other hand is learning how to take risks, you could go to Commitment and read *To Laugh is to Risk* (pg. 21).

Readiness assessment might determine that the group needs to think of why it is doing certain things, underlining the need to refocus. Looking through Commitment you could use *Butterflies in Formation* (pg. 13), or *Start by Doing* by St. Francis of Assissi (pg. 19). Or perhaps your readiness assessment has determined that there are safety issues to contend with that your group is not dealing with. So pull out Peter Willauer's *The Sea is Impersonal* (pg. 46), which is also lodged in the Community section. You might use this reading to stimulate a discussion around safety, thereby assessing the group's ability to deal with that issue along with preparing it for an activity that requires diligence.

Affect determinations might lead you to decide that the intense storming you're observing requires some lightening up or humor, like the *Insecure Camper* (pg. 57). On the other hand, the group may be too hyper and thus not able to focus on the task at hand. In the midst of such free flowing narcissism, some devaluing of members may be going on. Go to the Community section and use *I Corinthians:12* (pg. 26), where St. Paul speaks of the body not as a single organ, but many. "If one organ suffers, they all suffer together. If one flourishes, they all rejoice together."

Suppose the **Behavior** issue you observe involves stealing. Use *The Monkey Rope* (pg. 7) in the Commitment section. Carl Brown used it during an incident on Hurricane Island. He tied the group up with an anchor line in order to show them that they were all in the experience together. Another example of **Behavior** you may want to confront is an attitude toward the environment. Go to the Values section and use *Disposable Society* (pg. 113). It connects throwing away plastic silverware to disposing of relationships. Or you might want to get the group's attention with an especially hard hitting, bitter refrain about how modern society has treated the Great Plains environment. So utilize *The Great Plains* (pg. 126), also in the Values section.

Body issues that you observe could be represented by what is these days called burnout. This wasted person effect can be the result of many things, one of which is an over abundance of caring, represented by a co-dependent approach to the world. Symptoms can include a general lack of regard for self signalled by stress, fatigue, being out of shape, non-productive, unhappy, cranky, etc. Read *Burned Out?* by Edward Abbey (pg. 72) in the Journey section, where he sees himself as a "part-time Crusader." Or you could confront the soft belly of out of shape young people by reading *You Had To Be Taught* (pg. 131), in the Values section. It takes the pressure off of them by pointing out the fact that values regarding sugar, smoking, and "junk that clutters up your life..." were not created by them but are, on the other hand, taught by a culture that is in intense competition for their dollars and their souls.

It may be appropriate to use the Community category while a group's **Stage** of development is Storming. M. Scott Peck's definition would perhaps be appropriate, "a community is a group that can fight gracefully." You may come to the Transforming stage where the group has experienced intense closeness and is ready to terminate. An ideal reading is the *A.E.F. Doughboy Prayer* (pg. 110), that has as its supreme desire "to be radiant, to radiate life." That prayer is a great way to help the group members say goodbye to each other and to set goals for the future in the process.

Our intention here is not to offer a map or blueprint of the nuggets. Rather, we hope that the GRABBS process and the examples provided will help you get a feel for your own nugget selecting. As you get to know the nuggets you will be able to move them around and apply them to almost any situation.

Connected to your GRABBS selection is another critical "nugget factor" (contributed by Peter Coburn): "It is important to internalize them, to make the reading your own. Willen Lange can read *Birches* by Frost as though he wrote it, so much is the poem a part of him. Same with Bob Rheault and *Ulysses*. I have felt great incongruence in hearing imitators who have not yet made the reading their own. As the Skin Horse says, 'By the time you become Real most of you has been loved off, your eyes fall out and you get loose in the joints and very shabby.' Like my attempt from memory of this well-loved reading." Peter wanted to be clear that people should not be intimidated by the need to internalize, for that is a process in and of itself and requires practice. But we are tapping into a powerful oral tradition when we use readings, one that implies conviction, passion and connection with the material.

Happy Nuggetting!

A NOTE TO READERS

It has been our intention to track down the original source of all contributed material in order to request permission for its use and to give proper credit. We've spent nearly a full year in libraries, on the phone, writing letters and using any other methods possible to find the authors and copyright holders of the nuggets. We were not always successful. A number of nuggets are attributed to *author unknown,* and in some cases, *contributor unknown.*

Any material not attributed to the proper source, author or contributor is due to a lack of available information. It is not our intention to use or mis-use any material without requesting and giving proper credit. Anyone able to shed light on a work in this book that we have left unattributed or incorrectly attributed is requested to contact Project Adventure, Inc. so that in future editions the error or omission can be corrected.

Whenever possible the title associated with a nugget is derived from the work from which it was mined. However, for the majority of entries the titles are the editors' creations.

On Language

We are advocates of non-sexist language and have made efforts in our writing to utilize it. However, most of the writing in this book was produced before there was the present heightened consciousness regarding sexist issues and we cannot edit the writings of others. If you the reader choose to change them, you are certainly free to do so.

COMMITMENT

My Ship Is So Small
Ann Davison

I would have been here days ago, if only I'd had the courage of my convictions.

"You see," said my other self, a carping creature always ready to say I told you so. "If you would only learn to depend on yourself, and get it out of your head that a kind fate is standing by to pick you up when you fall, you would be all right. It is no good blaming the stars or luck or the lines in your hands, your problems are invariably of your own making."

I know that.

And only you can solve them. So you had better remember that in the future, and make an effort, instead of worrying around in concentric circles.

But I can't help worrying, I thought defensively. After all, I might be wrong.

Might be, sneered other self. Might be. You might be wrong about your landfall and sail onto a reef. You think how awful it would be to lose your ship and get into a tizzy about it, instead of thinking constructively how to avoid such a calamity...

Worrying is a form of running away. An escapism. A mental wringing of hands because of a refusal to face, not necessarily facts, but possiblities. When you see a possible consequence you don't like, you shy away. That's no good. You have to square up to 'em.

Courage of your convictions, I thought, which is where I came in. Courage — why, that's it, of course. That's the answer. That's what I've been looking for. How surprising, but how obvious when you see it.

Only you didn't see it, other self pointed out. You've been hanging by your teeth all these years because you confused courage with the conquering of physical fear. What you need is not the sort of courage that makes a man face danger. Criminals face hideous dangers sometimes, but they are the least courageous of all, for it takes courage to evaluate standards and live by them.

Then what is courage ?

An understanding and acceptance; but an acceptance without resignation, mark you, for courage is a fighting quality. It is the ability to make mistakes and profit by them, to fail and start again, to take heartaches, setbacks, and disappointments in your stride, to face every day of your life and every humdrum, trivial little detail of it and realize you don't amount to much, and accept the fact with equanimity, and not let it deter your efforts.

It is over now, I thought, stretching out on the bunk, at least the quest is.

Don't kid yourself, other self said sharply. You will go on muddling and flapping and floundering your way through life as you have always done.

But at least I'll know what is needed, even if I haven't it in me to use —

You've got it in you all right, everyone's got it in them. It isn't a special dispensation from a selective Providence. It is just a question of whether you have the guts to apply it.

Contributor: Celeste Archambault

Celeste comments: This reading is an excerpt from the book My Ship Is So Small by Ann Davison, about her single-handed voyage across the Atlantic in 1952 in the 23' sloop Felicity Ann. Ms. Davison was the first woman to sail singlehandedly across the Atlantic. She was middle aged, had lost her husband at sea a few years earlier, and was seeking a new direction in life. She writes on arriving in Nassau.

I have found this reading to be most powerful and valuable when read to students after they have attempted to meet a challenge, whether it be an individual or group initiative. Good examples are map and compass problems, ropes course initiatives, expeditions. Courage oftentimes relates to self-trusting in order to follow through — to make decisions and then to act upon them.

A time stands out in my memory when I used this reading with a group of young women doing an Outward Bound course in Maine. It was apparent that they had no "courage of their convictions" and were always willing to let the two young men watches in convoy with us make the final decisons about which course to sail, where to anchor and how best to solve a problem. They were coming from a place in themselves that said that what they thought didn't matter or wasn't important or right! My co-instructor and I decided to plan on overnight expedition for just them. As we left the dock we in-

COMMITMENT

structed them to anchor in a certain cove in the White Islands and explained that they were on their own from then on and would get no help from us whatsoever. Eyebrows raised as well as self-doubts and off we went with the students in command. In good time they masterfully guided the boat to the exact location we had designated, got ready to drop anchor, looked at me and asked, "Is this the place?" My silent reply shot terror and doubt in them. Without a moment's hesitation the anchor was stowed and we were off in search of their just-arrived-at destination. For hours they rowed in and out and all around this small chain of islands looking for this very same spot. At 2:45 AM we were again in position to drop anchor and again they looked at me and asked "Is this it? It has to be it!" Another silent reply…sighs…fatigue had set in and I heard the anchor chain slipping over the gunnel to her final destination at long last. The crew dropped to the deck, eyes blurry and wet. I read this passage and said goodnight.

Here are some questions to facilitate self-understanding.

• How does not trusting yourself affect your ability to make decisions and to act upon them?

• What goes on inside of you when you are faced with a challenge? Do you feel intimidated, scared or self-assured? Do you jump right in and try or do you hold back and question your ability to succeed?

• Imagine being in a boat by yourself crossing the Atlantic as Ann Davison was, having to solve all of your problems alone with no one else to rely upon. What problems do you now have that you have been putting off and waiting for someone else to solve for you?

Commitment

W.H. Murray

Until one is committed, there is some hesitancy, the chance to draw back, always ineffectiveness. Concerning all acts of initiative (and creation) there is one elementary truth, the ignorance of which calls countless ideas and splendid plans; that the moment one definitely commits oneself, providence moves too. All sorts of things occur to help one that would have never occurred. A whole stream of events, issues from the decision, raising in one's favor all manner of unforeseen incidents and meetings and material assistance, which no man could have dreamt would have come his way. I have learned a deep respect for one of Goethe's couplets:

"Whatever you can do, or dream you can, begin it. Boldness has genius, power and magic in it."

Contributor: Rafe Parker

Rafe's Comments: So many people never take a step into the unknown. So few people understand the magic that comes with making that first bold move. I have used this reading on countless courses since Willi Unsoeld introduced it to me on a winter course in the White Mountains. (Was it the winter of 1969?) The reading has always been helpful to me — it has also gotten me into a few scrapes! I have no doubt Willi was following Murray's philosophy right to the end.

Go for the Perfect Try

Sarah Smeltzer and Joe Petriccione

Growth is a never-ending process
that can be accomplished
under the most adverse circumstances.
Growth can be achieved
from one's attempts to...
"Go for the perfect try."

Contributor: Sarah Smeltzer

Sarah's comments: The group was struggling with personal goals on the last full day of an ABC workshop. Several people had issues around needing to be perfect...needing to achieve goals fully...setting strenuous personal standards. We talked about success, and how we are so programmed to crave victory and to fear failure. We talked about the alternative of giving our best to whatever we do, and measuring success by the degree to which we are at peace with ourselves.

The Hickory Jump was next. Joe chose to jump first from the furthest stump, from which he doubted he could succeed in reaching the trapeze. As he gathered himself for the leap, Sarah said, "Go for the perfect try." Joe's entire reality shifted. His success was his single, excellent leap. The trapeze was of no significance anymore. He missed it, but he walked away with the deep satisfaction of having gone for a perfect try.

To Build a Wooden Boat

Jim Schoel

Each plank, beveled
and butted in its climb
from garboard to sheer,
must be unique
and true as can be.
I go deep to find this, and take assurance
from my acts.
But this boat is not a house...
every piece done on time
does not assure the next.
There are few right angle plumb lines here.
Plywood and shingles cannot cover your errors.
Your creation will do battle with water,
not just shed it.
So it must come together
each day, new, complex,
each piece depending on the last, a figure in itself
demanding only that it
be right.
It asks nothing more,
is totally doable,
does not depend on sucking you dry.
You just do it
or you don't.

Spring, 1972

Hanging in There (an unpublished book of poems)

Contributor: Jim Schoel

Jim's comments: This poem came from a boatbuilding class I taught as part of the Sea as Teacher *action seminar (see the Project Adventure publication* Teaching Through Adventure*). I have read it at subsequent junctures where the group was dealing with issues of quality and choice. It has also been helpful to read it during community service activities, like tutoring projects, or repair and clean-up jobs at our camp site and ropes course. True teamwork is reflected in everyone playing a part, and fitting together in some way in order to get a job done. And it is true — you just do it, or you don't. The finished (or unfinished) job is the measure.*

Heads and Lives
Phil Salzman

It's easy to get things into our heads, but it's hard to get them into our lives.

Contributor: Phil Salzman

Phil's comments: I use this to communicate the difference between knowing and doing (most education feeds the mind to produce knowledge, rather than providing experience to produce understanding), and that leads to an awareness of "right practice" (Zen behavioral attitude), or what we ought to do. Real growth occurs when new and appropriate behavior develops from this awareness. There are so many examples of this and students can offer them up. A spin-off can be personal goal setting...or discussion about what it takes to "walk our talk."

In These Plethoric Times
H.G. Wells

But in these plethoric times when there is too much coarse stuff for everybody and the struggle for life takes the form of competitive advertisement and the effort to fill your neighbor's eye, there is no urgent demand either for personal courage, sound nerves or stark beauty, we find ourselves by accident. Always before those times the bulk of the people did not overeat themselves, because they couldn't, whether they wanted to or not, and all but a very few were kept fit by unavoidable exercise and personal danger. Now if only he pitch his standard low enough and keep free enough from pride, almost everyone can achieve a sort of excess. You can go through contemporary life fudging and evading, indulging and slacking, never really hungry nor frightened nor passionately stirred, your highest moment a mere sentimental orgasm, and your first real contact with primary and elemental necessities the sweat of your deathbed.

Contributor: Rafe Parker

Rafe's comments: This was read by Mike Isbell at Hurricane Island on July 19, 1971 and it has never been far from my consciousness. It has proved especially effective with adults, for reasons I am unsure about. There is an aggressive reprimandive quality about it that gets you to sit up in your chair and makes you smile somewhat nervously!

The Monkey Rope

Herman Melville

In the tumultuous business of cutting-in and attending to a whale, there is much running backwards and forwards among the crew. Now hands are wanted here, and then again hands are wanted there. There is no staying in any one place; for at one and the same time everything has to be done everywhere. It is much the same with him who endeavors the description of the scene. We must now retrace our way a little. It was mentioned that upon first breaking ground in the whale's back, the blubber-hook was inserted into the original hole there cut by the spades of the mates. But how did so clumsy and weighty a mass as that same hook get fixed in the hole? It was inserted there by my particular friend Queequeg, whose duty it was, as harpooneer, to descend upon the monster's back for the special purpose referred to. But in very many cases, circumstances require that the harpooneer shall remain on the whale till the whole flensing or stripping operation is concluded. The whale, be it observed, lies almost entirely submerged, excepting the immediate parts operated upon. So down there, some ten feet below the level of the deck, the poor harpooneer flounders about, half on the whale and half in the water, as the vast mass revolves like a tread-mill beneath him. On the occasion in question, Queequeg figured in the Highland costume — a shirt and socks — in which to my eyes, at least, he appeared to uncommon advantage; and no one had a better chance to observe him, as will presently be seen.

Being the savage's bowsman, that is, the person who pulled the bow-oar in his boat (the second one from forward), it was my cheerful duty to attend upon him while taking that hard-scrabble scramble upon the dead whale's back. You have seen Italian organ-boys holding a dancing-ape by a long cord. Just so, from the ship's steep side, did I hold Queequeg down there in the sea, by what is technically called in the fishery a monkey-rope, attached to a strong strip of canvas belted round his waist.

It was a humorously perilous business for both of us. For, before we proceed further, it must be said that the monkey-rope was fast at both ends; fast to Queequeg's broad canvas belt, and fast to my narrow leather one. So that for better or for worse, we two, for the time, were wedded; and should poor Queequeg sink to rise no more, then both usage and honor demanded, that instead of cutting the cord, it should drag me down in his wake. So, then, an elongated Siamese ligature united us. Queequeg was my own inseparable twin brother; nor could I any way get rid of the dangerous liabilities which the hempen bond entailed.

So strongly and metaphysically did I conceive of my situation then, what while earnestly watching his motions, seemed distinctly to perceive that my own individuality was now merged in a joint stock company of two: that my free will had received a mortal wound; and that another's mistake or misfortune might plunge innocent me into unmerited disaster and death. Therefore, I saw that here was a sort of interregnum in Providence; for its even-handed equity never could have sanctioned so gross an injustice. And yet still further pondering, I say, I saw that this situation of mine was the precise situation of every mortal that breathes; only, in most cases, he, one way or other, has this Siamese connexion with a plurality of other mortals. If your banker breaks, you snap; if your apothecary by mistake sends you poison in your pills, you die. True, you may say that, by exceeding caution, you may possibly escape these and the multitudinous other evil chances of life. But handle Queequeg's monkey-rope heedfully as I would, sometimes he jerked it so, that I came very near sliding overboard. Nor could I possibly forget that, do what I would, I only had the management of one end of it.

I have hinted that I would often jerk poor Queequeg from between the whale and the ship — where he would occasionally fall, from the incessant rolling and swaying of both. But this was not the only jamming jeopardy he was exposed to. Unappalled by the massacre made upon them during the night, the sharks now freshly and more keenly allured by the before pent blood which began to flow from the carcass — the rabid creatures swarmed round it like bees in a beehive.

And right in among those sharks was Queequeg; who often pushed them aside with his floundering feet. A thing altogether incredible were it not that attracted by such prey as a dead whale, the otherwise miscellaneously carnivorous shark will seldom touch a man.

Nevertheless, it may well be believed that since they have such a ravenous finger in the pie, it is deemed but wise to look sharp to them. Accordingly, besides the monkey-rope, with which I now and then jerked the poor fellow from too close a vicinity to the maw of what seemed a peculiarly ferocious shark — he was provided with still another protection. Suspended over the side in one of the stages, Tashtego and Daggoo continually flourished over his head a couple of keen whale-spades, herewith they slaughtered as many sharks as they could reach. This procedure of theirs, to be sure, was very disinterested and benevolent of them. They meant Queequeg's best happiness, I admit; but in their hasty zeal to befriend him, and from the circumstance that both he and the sharks were at times half hidden by the blood-muddied water, those indiscreet spades of theirs would come nearer amputating a leg than a tail. But poor Queequeg, I suppose, straining and gasping there with that great iron hook — poor Queequeg, I suppose, only prayed to his Yojo, and gave up his life into the hands of his gods.

Well, well, my dear comrade and twin brother, thought I, as I drew in and then slacked off the rope to every swell of the sea — what matters it, after all? Are you not the precious image if each and all of us men in this whaling world? That unsounded ocean you gasp in, is Life; those sharks, our foes; those spades, our friends; and what between sharks and spades you are in a sad pickle and peril, poor lad.

Moby Dick

Contributor: Carl Brown

Carl's comments: There had been a rash of thefts on Hurricane Island. Cases of peanut butter had gone missing. A person or persons within the Outward Bound community was not keeping faith with their fellows. At the morning meeting the thefts were announced, but no one came forward. That moment of silence was touched with the quiet of conspiracy. An instructor decided to rope his watch together in a long line, to stay that way until someone confessed participation or knowledge of the theft. By lunch nearly every watch on the island was roped together wearing on their faces the shameful expressions of the punished. I remember my students peering at me from the corners of their eyes, wondering if they would be next.

It seemed to me that the metaphor of the rope was being mis-used. That its meaning of our own mutual connectedness was coming unfrayed. I wanted to find an active way to splice it back.

That afternoon, I took my watch out in our pulling boat and passed out blindfolds to all but the bow watch and challenged them to sail down Hurricane Sound to Nun 8. The bow-watch was only allowed to indicate danger or success. At first they all thought that it was a crazy, impossible idea. But that being the Outward Bound way, they determined to give it a try. I had no idea what would happen nor was I prepared for what I witnessed. They began taking up jobs and telling the others where they were and what they were doing. One student took over the luff of the sail, feeling it with his fingers for wrinkles, luffing and fullness. Sailing terminology was making sense. Communication and working together was making the boat move. Now that they were sailing the next step was to sort out direction; where we were and where we were to go. They remembered the weather report. The wind was southwest. So they put the wind behind them and headed north toward the nun. They talked and visualized over the chart that they all had to piece together in their collective minds. They decided that the nun must bear more east. They needed another point of reference. The sun. There came a sight that remains my most vivid in all my years in Outdoor education. A group of young teenage boys with outstretched hands, feeling the air for the warmth of the summer's setting sun. Reaching some illusive, tactile consensus, they changed course. Nun 8 was not reached on the mark, yet 30 yards within a mile was not bad. More important, we had discovered something about ourselves.

When we returned to shore, the whole watch was excited and charged. Then I produced the rope. The rope was an anchor line and when I tied them in a single file the last man had to carry the anchor to which it was attached. Their jubilation over the afternoon's accomplishment instantly turned into mass dejection as we marched in this manner to the mess hall for dinner. At the mess hall all the other watches had been released from their long lines. Our arrival, roped up with the last man carrying an anchor was met with laughter and ridicule. My watch felt confused and ashamed. Getting permission to come a little late for dinner, I sat down with my watch. I explained to them that they were tied together this way not because I was punishing them, but because I wanted to make a point. I had for them a reading that I hoped, along with what they had achieved that afternoon, would help to illustrate what I had in mind. It was the Monkey Rope chapter. When I finished the looks of anger and shame had changed to expressions filled with a warm and salty understanding. I told them they could untie from the rope, but they refused. They wanted to dine that way. To take hold of the metaphor of connectedness and return it to its proper meaning.

In the mess hall the laughs and snickers quickly faded away before the dense sense of pride the watch carried into the room. The peanut butter was returned, all jars unopened and intact. It had been taken by a member of my watch who confessed the next day to the watch. It was the hardest thing that any of us had done the whole course which is why I gave him his pin.

I never could figure out why he never ate any of that peanut butter. I think perhaps he was trying to tell us something.

1855 Environmental Statement
Chief Seattle

Working in a most unscientific manner with nothing but intuition and love to guide him in the interpretation of his random data, the chief of the Duwamish Indians wrote president Franklin Pierce 130 years ago. "Savage" Chief Sealth (from whose name came the present city of Seattle) presented an environmental impact statement in 1855 which embodied the basic ecological insight — all things are connected. "Whatever befalls earth befalls the sons of the earth."

We know that the white man does not understand our ways. One portion of the land is the same to him as the next, for he is a stranger who comes in the night and takes from the land whatever he needs. The earth is not his brother but his enemy, and when he has conquered it, he moves on. He leaves his fathers' graves, and his children's birthright is forgotten. The sight of your cities pains the eye of the redman. But perhaps because the redman is a savage and does not understand.

There is no quiet place in the white man's cities. No place to hear the leaves of spring or the rustle of insects' wings. But perhaps because I am savage and do not understand, the clatter only seems to insult the ears. The Indian prefers the soft sound of the wind darting over the face of the pond, and the smell of the wind itself cleansed by a midday rain, or scented with a pinion pine. The air is precious to the redman. For all things share the same breath — the beasts, the trees, the man. The white man does not seem to notice the air he breathes. Like a man dying for many days, he is numb to the stench.

What is man without the beasts? If all the beasts were gone, man would die from great loneliness of spirit, for whatever happens to the beasts also happens to man. All things are connected. Whatever befalls the earth befalls the sons of the earth.

It matters little where we pass the rest of our days; they are not many. A few more hours, a few more winters, and none of the children of the great tribes that once lived on this earth, or that roamed in small bands in the woods, will be left to mourn the graves of a people once as powerful and hopeful as yours.

The Whites too shall pass — perhaps sooner than other tribes. Continue to contaminate your bed, and you will one night suffocate in your own waste. When the buffalos are all slaughtered, the wild horses all tamed, the secret corners of the forest heavy with the scent of many men, and the view of the ripe hills blotted by talking wires, where is the thicket? Gone. Where is the eagle? Gone. And what is to say goodbye to the swift and the hunt, the end of living and the beginning of survival? We might understand if we knew what it was that the white man dreams, what hopes he describes to his children on the long winter nights, what visions he burns into their minds, so they will wish for tomorrow. But we are savages. The white man's dreams are hidden from us.

The deer, the horse, the great eagle, these are our brothers, the earth is our mother...All things are connected like the blood which unites one family. Whatever befalls the earth befalls the sons (and daughters) of the earth.

Contributor: Jim Schoel

Jim's comments: When I first saw that Chief Seattle had made an important speech, I really got excited. That's because when I was a kid growing up in the Northwest (a small community south of Tacoma called Lakewood), I would listen to the radio in the morning. There was a local historian who always had a tale to relate: his name was Nard Jones, and he spoke many times about Sealth. So I grew up with a strong sense of his dignity and qualities as a person, and also of the compromising and tragic situation his people had been forced into.

When I look back over the times I've used this reading, I think of Ram Island in Gloucester, Massachusetts, near my home. It is a marsh island in the Annisquam River, with deep water moorage, some nice trees for a ropes course, a place to camp, a house for caretakers, and a wonderful sense of the wilderness, all a few miles from the city. We have been able to use it as an environmental and outdoor study center for all kinds of groups over the years. Often I would have groups of Gloucester school children out there, and after we had gotten all the cleanup accomplished and were ready to move on to our next activity (rowing dorys, or marsh study, rock climbing or the ropes course), I would pull out Chief Seattle and relate it to the Ram Island area. The sacredness of the land connected to individual courage was the central message I tried to convey. To illustrate this I spoke about a friend named George Gleason who had been a clammer in the area for many years and had almost singlehandedly saved Ram Island and the local marshes from development. George didn't have any degrees in environmental studies, but he did have clarity of thinking, the ability to defend his position, and courage. Fifteen years later, when the owner donated the land to the city, he expressed his appreciation to George for standing up the way he did, for he had learned that the property had much more than simply financial value.

An Artist's Failure
Thomas Hart Benton

The only way an artist can ever personally be a failure is to quit work.

Contributor: Jim Schoel

The Artist
Henry Miller

In a profound sense every great artist is hastening the end. A great artist is not simply a revolutionary, in style, form or content, but a rebel against the society he is born into. It is the artist who has the courage to go against the crowd; he is the unrecognized "hero of our time" and of all time.

When I Reach For My Revolver

Contributor: Todd Tinkham

At No Time in Life
Maggie Kuhn

At no time in life is there a lack of opportunity to rebuild and regroup.

Contributor: Jim Schoel

Bounders
Mike Stratton

The immediate challenge of Bounders is physical. The long-range effects have to do with the heart and the spirit. The course will not be easy! In fact, we provide some discomfort, fear, tension and lots of just plain hard work. But, one way or the other, it will be an important experience you won't forget for a long time to come.

Butterflies in Formation
Author Unknown

Remember, the trick is not getting rid of the butterflies in your tummy, it's getting them to fly in formation.

Contributor: Paul Radcliffe

Choosing to Function
Leonard Zunin

I believe that courage is all too often mistakenly seen as the absence of fear. If you descend by rope from a cliff and are not fearful to some degree, you are either crazy or unaware. Courage is seeing your fear, in a realistic perspective, defining it, considering alternatives and choosing to function in spite of risks.

Contact: The First Four Minutes

Contributor: Mike Stratton

Courage
Mark Twain

Courage is resistance to fear, mastery of fear — not absence of fear. Except a creature be part coward it is not a compliment to say it is brave...

Puddin'head Wilson

Contributor: Jim Schoel

Courage is the Complement
Robert Heinlein

Courage is the complement of fear. A person who is fearless cannot be courageous.

The Notebooks of Lazarus Long

Contributor: Karl Rohnke

Desert Pete
Tim Hansel

A man was walking across the desert, stumbling, almost dying of thirst, when he saw a well. As he approached the well, he found a note in a can close by. The note read: "Dear friend, there is enough water in this well, enough for all, but sometimes the leather washer gets dried up and you have to prime the pump. Now if you look underneath the rock just west of the well, you will find a bottle full of water, corked. Please don't drink the water. What you've got to do is take the bottle of water and pour the first half very slowly into the well to loosen up the leather washer. Then pour the rest in very fast and pump like crazy! You will get water. The well has never run dry. Have faith. And when you're done, don't forget to put the note back, fill up the bottle and put it back under the rock. Good luck. Have a fun trip. Sincerely, your friend, Desert Pete."

What would you do? You're on the verge of expiring from lack of water, and in reality, the bottle of water is only enough to quench your thirst, not save your life. Would you have the courage to risk it all?

This story is a powerful allegory about some of the essential ingredients in the Christian faith. First, there is evidence — there is a written message, the can with the letter in it, the bottle underneath the rock. Everything is in order, but there is no proof that you can really trust Desert Pete. The second element is risk. Here is a man dying of thirst asked to pour the only water he is sure of down the well. Faith is always costly. The third element is work. Some people have mistakenly interpreted faith as a substitute for work...Faith is not laziness. Desert Pete reminds us that after we trust and risk we must pump like crazy!

Holy Sweat

Contributor: Susan Schaefer

I Am Enamoured
Walt Whitman

I am enamoured of growing out-doors,
Of men that live among cattle, or taste of the ocean or woods,
Of the builders and steerers of ships, and the wielders of axes and mauls, and the drivers of horses;
I can eat and sleep with them week in and week out.
What is commonest, cheapest, nearest, easiest, is Me;
Me going in for my chances, spending for vast returns;
Adorning myself to bestow myself on the first that will take me;
Not asking the sky to come down to my good will;
Scattering it freely forever.

Leaves of Grass, Song of Myself

Contributor: Jim Schoel

The Incentive to Push On Further
Dag Hammarskjöld

Never let success hide its emptiness from you, achievement its nothingness, toil its desolation. And to keep alive the incentive to push on further, that pain in the soul which drives us beyond ourselves. Whither? That I don't know. That I don't ask to know.

Markings

Contributor: Jim Schoel

Life Yields Only
Dag Hammarskjöld

Life yields only to the conqueror. Never accept what can be gained by giving in. You will be living off stolen goods, and your muscles will atrophy.

Contributor: Mike Stratton

Mad Ones
Jack Kerouac

The only people for me are the mad ones, the ones who are mad to live, mad to talk, mad to be saved, desirous of everything at the same time, the ones who never yawn or say a commonplace thing, but burn, burn, burn like fabulous yellow roman candles exploding like spiders across the stars and in the middle you see the blue centerlight pop and everybody goes "Awww!"

On the Road

Contributor: Todd Tinkham

Man is Tough
William Faulkner

Man is tough. Nothing — war, grief, hopelessness, despair — can last as long as man himself can last; man himself will prevail over all his anguishes, provided he will make the effort to stand erect on his own feet by believing in hope and in his own toughness and endurance.

Contributor: Jim Schoel

Man's Influence
Robert F. Kennedy, 1966

Each time a man stands up for an ideal, or acts to improve the lot of the rest or strikes out against injustice, he sends forth a tiny ripple of hope, and crossing each other from a million different centers of energy and daring, those ripples build a current that can sweep down the mightiest walls of oppression and resistance...Few are willing to brave the disapproval of their fellows, the censure of their colleagues, the wrath of their society. Moral courage is a rarer commodity than bravery in battle or great intelligence. Yet it is the one essential, vital quality for those who seek to change a world that yields most painfully to change.

Contributor: Mike Stratton

Many Would Say

Anonymous

Many would say, "I'm afraid," if they had enough courage.

Contributor: Karl Rohnke

McMurphy

Ken Kesey

"By golly, he might do it," Cheswick mutters.

"Sure, maybe he'll talk it off the floor," Frederickson says.

"More likely he'll acquire a beautiful hernia," Harding says.

"Come now, McMurphy, quit acting like a fool; there's no man can lift that thing."

"Stand back sissies, you're using my oxygen."

McMurphy shifts his feet a few times to get a good stance, and wipes his hands on his thighs again, then leans down and gets hold of the levers on each side of the panel. When he goes to straining, the guys go to hooting and kidding him. He turns loose and straightens up and shifts his feet around again.

"Giving up?" Frederickson grins.

"Just limbering up. Here goes the real effort" — and grabs those levers again.

And suddenly nobody's hooting at him anymore. His arms commence to swell, and the veins squeeze up to the surface. He clinches his eyes, and his lips draw away from his teeth. His head leans back, and tendons stand out like coiled ropes running from his heaving neck down both arms to his hands. His whole body shakes with the strain as he tries to lift something he knows he can't lift, something everybody knows he can't lift.

But, for just a second, when we hear the cement grind at our feet, we think, by golly, he might do it.

Then his breath explodes out of him, and he falls back limp against the wall. There's blood on the levers where he tore his hands. He pants for a minute against the wall with his eyes shut. There's no sound but his scraping breath; nobody's saying a thing....

He stops at the door and looks back at everybody standing around.

"But I tried, though," he says. "Goddammit, I sure as hell did that much, now, didn't I?"

One Flew Over the Cuckoo's Nest

Contributor: Paul Radcliffe

Men Wanted

Sir Ernest Shackleton

Men wanted for hazardous journey, small wages, bitter cold, long months of complete darkness, constant danger, safe return doubtful, honor and recognition in case of success.

1906

Contributor: Mike Stratton

Outward Bound Ships' Articles

Captain Larry Bailey

Being of sound mind and body, I hereby sign on for the voyage hereto to live in good fellowship with my shipmates and to refrain from the use of tobacco, alcohol, and drugs for the extent of said voyage, and to perform all duties asked of me to the best of my ability...so help me God.

Contributor: Mike Stratton

Plant Your Own Garden

Author Unknown

So plant your own garden and decorate your own soul, instead of waiting for someone to bring you flowers.

Contributor: Jim Schoel

The Second Nut

Richard McKenna

The engine fought back. The coupling nuts would not come loose. There were thirty of them, ten to a flange, each nut four inches across, and they were welded in their threads by the rust of fifty years. Po-han held the wrench steady and Holman swung the twenty-pound sledge until his wrists felt wooden and his fingers trembled and he could hardly close his hands. In two hours, they got one nut off.

There was an art to sledging. Amateurs used a full-arm swing, and it was mostly noise and show. The best way was to move the sledge only about a foot, arms rigid and your right hand only a few inches from the hammer head, and you swing your whole body from the ankles. You made your whole body into a battering ram with the sledge as striking point, and you poured the fused momentum of bone, muscle and steel into what you hit. If what you hit did not yield, all the energy reflected back into you, and it jarred you to your heels.

The nuts would not yield. They tried each nut in turn, jacking the shaft to get at the lower ones, and stopped to ball-peen again and drip more kerosene, and sledged over the nuts a second time. Holman sledged with an increasingly desperate blasting anger, and with each blow he could feel in his right hand the back jar of unavailing force re-enter him. Steel on steel struck sparks and added a sulfurous flavor to the steamy, damp, kerosene smell. Both men dripped sweat. By suppertime they had loosened one more nut. "Ain't much, for a day's work," Holman said wearily. Po-han grinned. He was happy about that second nut.

The Sand Pebbles

Contributor: Peter Coburn

Start By Doing
Saint Francis

Start by doing what's necessary, then what's possible and suddenly you are doing the impossible.

Contributor: Scott Garman

The Rules of the Game
John Ciardi

Every game that was ever invented by man consists in making the rules harder for the fun of it.

Contributor: Mike Stratton

The Tightrope Walker
Tim Hansel

Once there was a tightrope walker who performed unbelievable aerial feats. All over Paris, he had done tightrope acts at great heights. He followed his initial acts with succeeding ones, while pushing a wheelbarrow. A promoter in America heard about this and wrote him, inviting the daredevil to perform his act over the waters and dangers of Niagara Falls. He added, "I don't believe you can do it." The tightrope walker accepted the challenge. After much promotion and planning, the man appeared before a huge crowd gathered to see the event. He was to start on the Canadian side and walk to the American side. Drums rolled and everyone gasped as they watched the performer walk across the wire blindfolded with a wheelbarrow. When he stepped off on the American side, the crowd went wild. Then the tightrope walker turned to the promoter and said, "Well, now do you believe I can do it?"

"Sure I do," the promoter answered. "I just saw you do it."

"No, no, no," said the tightrope walker. "Do you really believe I can do it?"

"I just said I did."

"I mean do you really believe?"

"Yes, I believe!"

"Good," said the tightrope walker, "then get in the wheelbarrow and we'll go back to the other side."

Holy Sweat

Contributor: Rich Obenschain

To Be Of Use

Marge Piercy

The people I love the best
jump into work head first
without dallying in the shallows
and swim off with sure strokes almost out of sight.
They seem to become natives of that element,
the black sleek heads of seals
bouncing like half-submerged balls.

I love people who harness themselves, an ox to a heavy
   cart,
who pull like water buffalo, with massive patience,
who strain in the mud and the muck to move things
   forward,
who do what has to be done, again and again.

I want to be with people who submerge
in the task, who go into the fields to harvest
and work in a row and pass the bags along,
who stand in the line and haul in their places,
who are not parlor generals and field deserters
but move in a common rhythm
when the food must come in or the fire be put out.

Circles on the Water

Contributor: Celeste Archambault

To Laugh is to Risk
Anonymous

To laugh is to risk appearing the fool,
To weep is to risk appearing sentimental,
To reach out for another is to risk involvement,
To expose our feelings is to risk exposing our true self,
To place your ideas and dreams before the crowd is to
    risk loss,
To love is to risk not being loved in return,
To live is to risk dying,
To hope is to risk despair,
To try at all is to risk failure,
But risk we must, because the greatest hazard in life is to
risk nothing.
The man, the woman who risks nothing, does nothing,
    has nothing,
is nothing.

Contributor: George Ashley

Ultimate Measure
Martin Luther King, Jr.

The ultimate measure of a man is not where he stands in
moments of comfort and convenience, but where he stands
at times of challenge and controversy.

The Words of Martin Luther King, Jr.

Contributor: Paul Radcliffe

We Make a Fetish of Success
Steven Muller

Americans do not understand nor do they live well with
failure. Yet it is an inevitable part of the human condition:
no one can win them all. We have made a national fetish of
success and victory — I think to a dangerous degree.

I believe in achievement, but I believe the crucial factor is
the effort rather than the result. Who can do more than one's
best? Who can ask more than to give the most one has? A
successful person is one who is productive to the peak of his
capacity and who is comfortable with his own self.

This sounds so simple, but it is not commonly accepted.
Success is equated with wealth, power, prestige and notori-
ety. These are dubious assets — some crave and possess
them, but find no happiness or fulfillment in this applica-
tion. Clearly, they are not available to all. But a successful
life is possible for each of us. My argument is that as people
we try to shut out the realities of failure and of death (the

ultimate failure) and that this is unhealthy. Each of us will die and each of us will fail at things. Can't we admit that and live with it?

The message of organized Christianity has been that death is a natural climax of life, not a hateful evil. Failure is no disgrace. He who never fails can never have tried very hard, and how do we know our limits without failure?

Strive for the self-respect that comes from giving your best to whatever you do, and measure success by the degree to which you are at peace with yourself.

Accept failure as natural and unavoidable and do not allow the best that is in you to be stunted by fear of failure.

Admit your failures easily, not only because it is neurotic to deny them, but because to fail the first time may be the best way to learn how to master the problem the next time, and also because failure in one direction may be the only road to mastery in others.

Do not envy the trappings of success in others. There is so much less to them than meets the eye.

You will not, thank God, each and every one of you become President of the U.S. or U.S. Steel, or latter-day Elizabeth Taylors, Ella Fitzgeralds, Mickey Mantles, or John Paul Gettys. But to each of you I wish a successful life of self-respect based on your best efforts exerted without fear of failure.

From a speech to the midwinter graduating class of the University of Maryland. Printed in the Washington Post.

Contributor: Paul Radcliffe

When You Get Into a Tight Place

Harriet Beecher Stowe

When you get into a tight place and everything goes against you, till it seems as though you could not hang on a minute longer, never give up then, for that is just the place and time that the tide will turn.

Contributor: Mike Stratton

COMMUNITY

Beware of Me, My Friend
Author Unknown

My name is Cocaine — "Coke" for short —
And I first entered this country without a passport.
Ever since then, I've been hunted and sought
By junkies and pushers and plain-clothes "dicks" —
But mostly by users who need a "quick fix."
I'm more valued than diamonds, more treasured than
    gold.
My power is boundless once I take hold.
I'll take you to depths you never thought you would
    reach:
I'll make a preacher not want to preach.
I'll make a schoolboy forget all his books, I'll make a
    beauty queen neglect her looks.
I'll take a renowned speaker and make him a bore,
I'll take your mama and make her a whore.
I've taken all kinds of people under my wing.
Just look around you for the effects of my "sting."
I've got daughters turning on their mothers,
I've got sisters stealing from their brothers.
I've got burglars robbing the courthouse.
I've got husbands pimping their spouse.
I'm the king of Crime and Prince of Destruction,
And I'll cause your body organs not to function.
I'll cause your babies to be born hooked,
I'll turn an honest man into a Crook.
I'll make you rob, steal and kill.
When you're under my power, you have no will.
My name is "Big C" and that name is no lie.
Some call me the "White lady," but no "lady" am I —
I've destroyed actors, sports heroes and politicians.
I've reduced bank accounts to zero from millions.
I'm a bad habit, too tough for any man.
Think you can control me? You're a fool. No one can.
Yeh, I'm raising hell all over the earth.

You don't believe me? Before you pass judgement check
the "funny farms" first.

My puppets are standing on corners selling "rocks." Yeh,
I've got it made!

Shootings and stabbings are my stock and trade.

Well, now you know what I do.

So remember my friend, it's all up to you.

For once you decide to ride the "White Horse" of Cocaine
be advised:

You'd best ride me well,

For it's my intention to RIDE YOU TO HELL!

Contributor: Shelia Torbert

*Shelia's comments: During the past year I noticed a
lot of my peers were users of cocaine. Some were users
because it was free and they wanted to hang with the
crowd. Others were users because their organs needed it. I
began to see attitude changes in them. Some of them lost
their jobs, homes and family. I saw the changes in them. I
never said anything to them, because I felt that it was
their problem, not mine. Then it dawned on me who they
were — teachers, bus drivers, the unemployed and cops.
It became my problem because, not only were they killing
themselves, they were in charge of other people's lives
which they could also destroy. But what really got to me
was when I was taking a walk and an eleven-year-old kid
approached me to sell me some coke. That freaked me
out and then I saw other kids buying the coke from the
same eleven-year-old kid.*

*Later that day a friend came by. I talked to her about
what I saw and how I was feeling. She told me about how
she got to be a user and some of the things she went
through. And how lucky she was to have found this poem
because she was about ready to hit "rock bottom." She
said whenever she feels the need for a hit she reads this
poem. This poem means a great deal to her and to myself.
I can only hope and pray that you too will find a place for
this poem.*

*Editor's note: When Shelia read this poem to a group,
she also talked about coming home at night in her
neighborhood and seeing young kids shoot at each other
across the street around her car. "I could have been
killed!" She spoke with such feeling that tears filled up
her eyes. She encouraged all of us to give copies of it to
anyone, anywhere.*

Cyclops

Margaret Atwood

You are going along the path,
mosquito-doped, with no moon, the flashlight
a single orange eye

unable to see what is beyond
the capsule of your dim
sight, what shape

contracts to a heart
with terror, jumps
among the leaves, what makes
a bristling noise like a fur throat.

Is it true you do not wish to hurt them?

Is it true you have no fear?
Take off your shoes then,
let your eyes go bare,
swim in their darkness as in a river

do not disguise
yourself in armour.

They watch you from hiding:
you are a chemical
smell, a cold fire, you are
giant and indefinable

In their monstrous night
thick with possible claws
where danger is not knowing

you are the hugest monster.

Procedures for Underground

Contributor: James Raffan

Jim's comments: This is one of my favorite night hike dressings because it focuses attention on the land and tends to re-orient participant fear in a productive way. It reads well with a red flashlight in a darkened bus on the side of the road just before putting students in on their night hike.

From Group to Community

M. Scott Peck, M.D.

In genuine community there are no sides. It is not always
easy but by the time they reach community the members
have learned how to give up cliques and factions. They have
learned how to listen to each other and how not to reject
each other. Sometimes consensus in community is reached
with miraculous rapidity. But at other times it is arrived at
only after lengthy struggle. Just because it is a safe place
does not mean community is a place without conflict. It is,
however, a place where conflict can be resolved without
physical or emotional bloodshed and with wisdom as well
as grace. A community is a group that can fight gracefully.

The Different Drum: Community Making and Peace

Contributor: Mark Murray

*Mark's comments: An appropriate reading as the
group is nearing a stage in their development where they
are beginning to sense a transformation of themselves
from merely a group to something a little more special
and a lot more significant — a community.*

I Corinthians 12

Saint Paul

A body is not one single organ, but many. Suppose the
foot should say, "Because I am not a hand, I do not belong to
the body," it does belong to the body nonetheless. Suppose
the ear were to say, "Because I am not an eye, I do not
belong to the body," it still does belong to the body. If the
body were all eye, how could it hear? If the body were all
ear, how could it smell? But in fact, God appointed each
limb and organ to its own place in the body, as he chose. If
the whole were one single organ, there would not be a body
at all; in fact, however, there are many different organs, but
one body. The eye cannot say to the hand, "I do not need
you," nor the head to the feet, "I do not need you."

God has combined the various parts of the body, giving
special honor to the humbler parts, so that there might be no
sense of division in the body, but that all its organs might
feel the same concern for one another. If one organ suffers,
they all suffer together. If one flourishes, they all rejoice
together.

Paul's first letter to the Christians at Corinth

Contributor: Will Lange III

*Will's comments: Morning readings at Hurricane
Island were always held on the big slab of granite that
sloped down to the sea between the mess hall and the*

staff lounge. On clear days, the whole anchorage lay spread out beneath us, from the main pier to Valley Cove.

The diversity of the fleet was striking. There were peapods and dinghies, daysailers, outboard workboats and fast outboards, patrol boats, the Bertram rescue boat, the MV Hurricane, and, of course, the ranks of ruggedly beautiful pulling boats. Different as they were, all of them had essential roles in the operation of the island. Each was fitted for the particular part it played, and did nothing else nearly as well as what it was designed to do. Yet without even one of them doing the job it was best suited for, the whole program suffered.

Watches of young men and women are like that: a diversity of talents, a variety of gifts, combined in a single body that can succeed only if every part of it succeeds. Every watch has its daysailers, its workboats, its pulling boats, and, of course, its dinghies; and every one of them has unique gifts to offer the entire group. It is the group's job to find these talents in its members and help them to flourish.

The Learning Gate
Carol Cornwell

Do you like what you see
Every time you look at me?
  My full lips
    broad nose
    smooth, ebony skin
    crinkly hair
  My smile within?
My name may be
    Aukram
       or
    Imam
    Nia, Takia
    Nefertiti
       or
    Hassan
But whatever my name
  and the history it brings
How will you teach me
If you don't learn the
  rhythms I sing?
'Cause if you don't know
  what is special to me
How will I learn from 9 'till 3?

Original poem by a former Cambridge (Mass.) Rindge and Latin High School student

Contributor: Bill Toomey

Bill's comments: Every time I read the poem called The Learning Gate by Carol Cornwell, I relive the experience of looking into the eyes of a woman who after reading the poem, turned to me and said, "Do you know me?"

In February every year the Cambridge School Department and Project Adventure Inc., collaborate in offering an indoor UMPA (Urban Modification of Project Adventure) workshop. This is the third year I have been a co-leader for the workshop.

Cambridge has sophisticated ropes courses in all of its inner city school gyms, with both high and low elements. The purpose of the workshop is to introduce these elements, and the adventure process, to teachers who work in inner city environments, some that have no ropes courses. It was interesting that, to me, the people who attended the workshop were people who may have come from similar backgrounds to the very students who attend the school. I could feel the excitement as they expressed the thought of their students learning and growing using our elements.

During the lunch break, I noticed one of the participants reading the poem I speak of. It was hanging on the bulletin board and I explained that it had been written by a student and a copy had been sent to all the teachers in the system. The woman turned, made eye contact with me and said "Do you know what is special to me (in this poem)? It is the question, 'Do you know me?'" The question caught me so off-guard that I ignored it and assured her that I would get her a copy of the poem. She had her copy, but the rhetorical question still lingered; Do I know you? Do I know you as well as I profess to know you? Do I care to know you? Did I know you in my earlier years of teaching? My thoughts turned toward my students and the same questions ran through my mind. Is twenty years on the job too long? If I want to know you, how do I do that now?

People left the workshop, and the staff was pleased that the participants found it to be, by and large, a good learning experience. For me, too. On Monday I couldn't help wondering what was special to the black kid, the white kid, the Asian kid, and the Hispanic kid.

Twenty years teaching in an urban environment may be too long, and it may not, but the interaction with that teacher helped me to create some other kinds of activities for my students and a renewed perspective for my colleagues.

Thanks, _____ wherever you are.

Ministry of Bearing
Dietrich Bonhoeffer

The freedom of the other person includes all that we mean by a person's nature, individuality, endowment. It also includes his weaknesses and oddities, which are such a trial to our patience, everything that produces frictions, conflicts and collisions among us. To bear the burden of the other person means involvement with the created reality of the other, to accept and affirm it, and, in bearing with it, to break through to the point where we take joy in it.

Life Together

Contributor: Susan Schaefer

Susan's comments: I discovered this nugget during my first year as a "sherpa" on La Vida, a Christian wilderness program in the Adirondack Mountains. We were a young staff, still in college or just out, and eager to prove how tough we were. In our first weeks of training we found ourselves trying to erase insecurity by competing with one another. A breakthrough for the group came when people finally started admitting their fears and stopped judging themselves by the number of climbs they could do, or the type of job they were assigned, or which mountains they led their groups over. I used this reading during a morning devotional to point out that everyone has weaknesses of one kind or another. It was met with silence, but a few days later one of the sherpas confided in me that it had really made her think about her actions and attitude about others' weaknesses. Since then, we've all seemed to follow different paths, but the lesson is always with us. We just need to learn it over again with each new group we enter. When I began teaching junior high, I had to learn to accept the frustrations of adolescents and not take them personally, and to force myself to smile when I felt like screaming. For another sherpa who went on to become a forest ranger, it meant realizing that not all campers have the same respect for the woods, and learning the most effective ways to educate them. Along with Bonhoeffer's images of accepting, affirming, breaking through — I add my own image of embracing the person, conflict, etc.

Power of Experience
Mark Twain

I felt good and all washed clean of sin for the first time I had ever felt so in my life, and I knowed I could pray now. But I didn't do it straight off, but laid the paper down and set there thinking — thinking how good it was all this happened so, and how near I come to being lost and going to hell. And went on thinking. And got to thinking over our trip down the river; and I see Jim before me all the time: in the day and in the nighttime, sometimes moonlight, sometimes storms, and we a-floating along, talking and singing and laughing. But somehow I couldn't seem to strike no places to harden me against him, but only the other kind. I'd see him standing my watch on top of his'n, 'stead of calling me, so I could go on sleeping; and see him how glad he was when I come back out of the fog; and when I come to him again in the swamp, up there where the feud was; and suchlike times; and would always call me honey, and pet me, and do everything he could think of for me, and how good he always was; and at last I struck the time I saved him by telling the men we had smallpox aboard, and he was so grateful, and said I was the best friend old Jim ever had in the world, and the *only* one he's got now; and then I happened to look around and see that paper.

It was a close place. I took it up, and held it in my hand. I was a-trembling, because I'd got to decide, forever, betwixt two things, and I knowed it. I studied a minute, sort of holding my breath, and then says to myself: "All right then, I'll *go* to hell" — and tore it up.

The Adventures of Huckleberry Finn

Contributor: James Raffan

Jim's comments: After sailing down the Mississippi on a raft with Jim, a runaway slave, Huck Finn's conscience finally catches up with him, and compels him to do the socially acceptable thing and turn Jim in by writing a note to Miss Watson, his owner. This excerpt picks up Huck's thoughts immediately after writing the note. It's a powerful example of experience mediating a decision between what society thinks is right and what Huck feels in his heart. In the end, experience sways the balance.

I encountered this reading teaching with Joe Nold. He used it to wind up an intense 10-day residential course in outdoor and experiential education for student teachers. Although it worked well in that context, I prefer to use the piece to pull group thoughts out of the mud after a tough day on the river. It reads beautifully around the campfire and can flow into a discussion of how the physical and mental courage required to meet difficult challenges often metaphorically melds into a more universal kind of moral courage.

Beyond Words

Ouida

There may be moments in friendship, as in love, when silence is beyond words. The faults of our friend may be clear to us, but it is well to seem to shut our eyes to them. Friendship is usually treated by the majority of mankind as a tough and everlasting thing which will survive all manner of bad treatment. But this is an exceedingly great and foolish error; it may die in an hour of a single unwise word; its conditions of existence are that it would be dealt with delicately and tenderly, being as it is a sensitive plant and not a roadside thistle. We must not expect our friend to be above humanity.

Contributor: Jim Schoel

The Butterfly

Nikos Kazantzakis

I remembered one morning when I discovered a cocoon in the bark of a tree, just as the butterfly was making a hole in the case and preparing to come out. I waited a while, but it was too long appearing and I was impatient. I bent over it and breathed on it to warm it. I warmed it as fast as I could and the miracle began to happen before my eyes, faster than life. The case opened, the butterfly started slowly crawling out and I shall never forget my horror when I saw how its wings were folded back and crumpled; the wretched butterfly tried with its whole trembling body to unfold them. Bending over it, I tried to help with my breath. In vain. It needed to be hatched out patiently and the unfolding of the wings should be a gradual process in the sun. Now it was too late. My breath had forced the butterfly to appear, all crumpled, before its time. It struggled desperately, and, a few seconds later, died in the palm of my hand.

That little body is, I do believe, the greatest weight I have on my conscience. For I realize today that it is a mortal sin to violate the great laws of nature. We should not hurry, we should not be impatient, but we should confidently obey the eternal rhythm.

I sat on a rock to absorb this New Year's thought. Ah, if only that little butterfly could always flutter before me to show me the way.

Zorba the Greek

Contributor: Mike Stratton

Climb if You Will

Edward Whymper

Climb if you will, but remember that courage and strength are nothing without prudence, and that a moment's negligence may destroy the happiness of a lifetime.

Contributor: Ted Woodward

Editor's note: Whymper led the first expedition to conquer the summit of the Matterhorn in 1865. Four of his men were lost in the descent.

Concrete Jungle

Bob Marley and the Wailers

No sun will shine in my day today.
The high yellow moon won't come out to play.
I said darkness has covered my light
Has changed my day into night.
Where is the love to be found?

Someone tell me 'cause
Light must be somewhere to be found
Instead of a concrete jungle
Where the living is heartless
Concrete jungle
And man, you've got to do your best.

No chains around my feet
But I'm not free.
I know I am bound here in captivity
I've never known happiness,
Never known what sweetness is.
Still I'll be always laughing like a clown.
Won't someone help me cause
I've got to pick myself from off the ground
In this concrete jungle,
Concrete jungle.
Why won't you let me be now?

Contributor: Jim Schoel

Crazy Horse

Ian Frazier

Personally, I love Crazy Horse because even the most basic outline of his life shows how great he was; because he remained himself from the moment of his birth to the moment he died; because he knew exactly where he wanted to live, and never left; because he may have surrendered, but he was never defeated in battle; because, although he was killed, even the Army admitted he was never captured; because he was so free that he didn't know what a jail looked like; because at the most desperate moment of his life he only cut Little Big Man on the hand; because, unlike many people all over the world, when he met white men he was not diminished by the encounter; because his dislike of the oncoming civilization was prophetic; because the idea of becoming a farmer apparently never crossed his mind; because he didn't end up in the Dry Tortugas; because he never met the President; because he never rode on a train, slept in a boarding house, ate at a table; because he never wore a medal or a top hat or any other thing that white men gave him; because he made sure that his wife was safe before going to where he expected to die; because, although Indian agents, among themselves, sometimes referred to Red Cloud as Red and Spotted Tail as Spot, they never used a diminutive for him; because, deprived of freedom, power, occupation, culture, trapped in a situation where bravery was invisible, he was still brave; because he fought in self-defense, and took no one with him when he died; because, like the rings of Saturn, the carbon atom, and the underwater reef, he belonged to a category of phenomena that our technology had not then advanced far enough to photograph; because no photograph or painting or even sketch of him exists; because he is not the Indian on the nickel, the tobacco pouch, or the apple crate. Crazy Horse was a slim man of medium height with brown hair hanging beneath his waist and a scar above his lip. Now, in the mind of everyone who imagines him, he looks different.

I believe that when Crazy Horse was killed something more than a man's life was snuffed out. Once, America's size in the imagination was limitless. After Europeans settled and changed it, working from the coasts inland, its size in the imagination shrank. Like the center of a dying fire, the Great Plains held that original vision longest. Just as people finally came to the Great Plains and changed them, so they came to where Crazy Horse lived and killed him. Crazy Horse had the misfortune to live in a place that existed both in reality and in the dreams of people far away; he managed to leave both the real and the imaginary place unbetrayed. What I return to most often when I think of Crazy Horse is the fact that in the adjutant's office he refused to lie on the cot. Mortally wounded, frothing at the mouth, grinding his teeth in pain, he chose the floor instead. What a distance there is between that cot and the floor! On the cot, he would have been, in some sense, ours: an object of pity, and accident victim, the noble red man, the last of his race. But on the floor Crazy Horse was Crazy Horse still. On the floor, he began to hurt as the morphine wore off. On the floor, he remembered Agent Lee, summoned him, forgave him. On the floor, unable to rise, he was guarded by soldiers even then. On the floor, he said goodbye to his father and Touch the Clouds, the last of the thousands who once followed him. And on the floor, still as far from white men as the limitless continent they once dreamed of, he died. Touch the Clouds pulled the blanket over his face: "That is the lodge of Crazy Horse." Lying where he chose, Crazy Horse showed the rest of us where we are standing. With his body he demonstrated that the floor of an Army office was part of the land, and that the land was still his.

Crazy Horse was my gran'father!

The Great Plains

Contributor: Jim Schoel

Do You Think You Are a Mistake

Author Unknown

Do you think you are a mistake…just because you made one?

Contributor: Jim Schoel

Enviropimp

Randolph Bayliss

In Prince William Sound, the SW quadrant, including Naked, Knight, and Green Islands and mainlands to the West, the sounds and sights of death have now quieted. In the first fortnight, you could see birds and otters struggle to keep warm, to preen fur and feathers of oil. You could hear a chilled, muffled whine of despair. Now, silence dominates in a noticeable, eerie intensity. The whine of despair has now been displaced by the stench of death. The putrid odor of decaying life recalls the battlefields of unburied dead.

At the otter cleaning center, you can revive them from hypothermia just in time to see their livers and kidneys shut down.

It's the saddest thing I've seen. Sadder than war because the animals have innocence, because the poison of oil works so slowly, with such agony.

Humans do not deserve stewardship of the Earth, and at this poisonous rate, we won't be around to claim it.

From a letter to a friend, 1989

Contributor: Beau Bassett

Failure as Success

William George Jordan

Many of our failures sweep us to greater heights of success than we ever hoped for in our wildest dreams. Life is a successive unfolding of success from failure. In discovering America Columbus failed absolutely. His ingenious reasoning and experiment led him to believe that by sailing westward he would reach India. Every redman in America carries in his name "Indian" the perpetuation of the memory of the failure of Columbus. The Genoese navigator did not reach India; the cargo of "souvenirs" he took back to Spain to show to Ferdinand and Isabella as proofs of his success really attested his failure. But the discovery of America was a greater success than was any finding of a "back door" to India.

Our highest hopes are often destroyed to prepare us for better things. The failure of the caterpillar is the birth of the butterfly; the passing of the bud is the becoming of the rose; the death or destruction of the seed is the prelude to its resurrection as wheat. It is at night in the darkest hours, those preceding dawn, that plants grow best, that they most increase in size. May this not be one of Nature's gentle showings to man of the times when he grows best, of the darkness of failure that is evolving into the sunlight of success? Let us fear only the failure of not living the right as we see it, leaving the results to the guardianship of the Infinite.

There is no honest and true work, carried along with constant and sincere purpose, that ever really fails. If it sometimes seems to be wasted effort, it will prove to us a new lesson of "how" to walk; the secret of our failures will prove to us the inspiration of possible successes. Man, living with the highest aims as best he can, in continuous harmony with them, is a success, no matter what statistics of failure a near-sighted or half-blind world of critics and commentators may lay at his door.

The Majesty of Calmness

Contributor: Ann Smolowe

Friendship

Kahlil Gibran

And the youth said, Speak to us of Friendship.
And he answered saying;
Your friend is your needs answered.
He is your field which you sow with love and reap with thanksgiving.
And he is your board and your fireside.

For you come to him with your hunger, and seek him for
peace.

When your friend speaks his mind you fear not the "nay"
in your own mind, nor do you withhold the "ay."
And when he is silent your heart ceases not to listen to
his heart;
For without words, in friendship, all thoughts, all desires,
all expectations are born and shared, with joy that is
unacclaimed.
When you part from your friend, you grieve not;
For that which you love most in him may be clearer in his
absence, as the mountain to the climber is clearer from the
plain.

The Prophet

Contributor: Mike Stratton

Gates of Eternity
Harriet Beecher Stowe

In the Gates of eternity the black hand and the white
hand hold each other with an equal clasp.

Contributor: Jim Schoel

The Hundredth Monkey
Ken Kesey

The Japanese monkey has been observed in the wild for a
period of over 30 years. In 1952, on the island of Koshima,
scientists were providing monkeys with sweet potatoes
dropped in the sand. The monkeys liked the taste of the raw
sweet potatoes, but found the dirt unpleasant. An 18-month
old female named Imo found that she could solve the
problem by washing the potatoes in a nearby stream. She
taught this trick to her mother. Her playmates also learned
this new way and they taught their mothers, too.

This cultural innovation was gradually picked up by
various monkeys before the eyes of the scientists. Between
1952 and 1958, all the young monkeys learned to wash the
sandy sweet potatoes to make them more palatable. Only the
adults that imitated their children benefitted from this social
improvement. Other adults kept eating the dirty potatoes.

Then something startling took place. In the autumn of
1958, a certain number of Koshima monkeys were washing
sweet potatoes — the exact number is not known. Let us
suppose that when the sun rose one morning there were 99
monkeys on Koshima Island who had learned to wash their

sweet potatoes. Let's further suppose that later that morning, the hundredth monkey learned to wash potatoes.

Then it happened!

By that evening, almost everyone in the tribe was washing sweet potatoes before eating them. The added energy of this hundredth monkey somehow created an ideological breakthrough!

But notice. A most surprising thing observed by these scientists was that the habit of washing sweet potatoes then jumped over sea. Colonies of monkeys on other islands and the mainland troop of monkeys at Takasakiyama began washing their sweet potatoes!

Thus when a certain critical number achieves an awareness, the new awareness may be communicated from mind to mind. Although the exact number may vary, the Hundredth Monkey Phenomenon means that when only a limited number of people know of a new way, it may remain the conscious property of these people. But there is a point at which if only one more person tunes in to a new awareness, a field is strengthened so that this awareness is picked up by almost everyone!

New Perspective Journal

Contributor: Scott Garman

I Am Done With Great Things

William James

I am done with great things and big things, great institutions and big success, and I am for those tiny invisible molecular moral forces that work from individual to individual by creeping through the crannies of the world like so many rootlets, or like the capillary oozing of water, yet which, if you give them time, will rend the hardest monuments of man's pride.

Contributor: Mike Stratton

I Never Held Your Hand

Tim Churchard

You're leaving and we haven't met
I should have asked and now regret
I didn't take the time,
I didn't take the time.
You're the one I didn't love
I wanted to...to rise above
To walk alone by the sea

And get beyond our mediocrity.
How many people I don't know
How many chances I let go
Passing strangers...we might be friends
Even lovers...it never ends
The people I don't know.
It's just a song of loneliness
About friends and lovers I have missed
The touch of hands, heart and mind
And all the rhythms of us in time.
My spirit soars when with a friend
Please come back and I'll try again
To let you know just who I am
I want you to know who I am.
I never chanced to watch you cry
To hold your hand and wonder why
To be so close and touch your tears
To talk of essence and passing years.
Is it me that I don't know
While loving you I have to grow
So hold me close and love me too
And tell me how to love you
I want to know how to love you.
How many people I don't know
How many chances I let go
Passing strangers, we might be friends
Even lovers...it never ends
The people I don't know.

Contributor: Tim Churchard

If I Can Stop One Heart

Emily Dickinson

If I can stop one Heart from breaking
I shall not live in vain
If I can ease one Life the Aching,
Or cool one Pain;

Or help one fainting Robin
Unto his Nest again,
I shall not live in Vain.

Contributor: Jim Schoel

Listen With Your Ears

Author Unknown

When I ask you to listen to me and you start giving
    advice, you have not done what I asked.

When I ask you to listen to me and you begin to tell me
why I shouldn't feel that way, you are trampling on
my feelings.
When I ask you to listen to me and you feel you have to
do something to solve my problem, you have failed
me, strange as that may seem.
Listen! All I asked was that you listen, not talk or do —
just hear me.
Advice is cheap; fifteen cents will get you both Dear Abby
and Billy Graham in the same newspaper.
And I can do for myself. I am not helpless. Maybe
discouraged and faltering, but not helpless.
When you do something for me that I can and need to do
for myself, you contribute to my fear and inadequacy.
But, when you accept as a simple fact that I do feel what I
feel, no matter how irrational, then I can quit trying to
convince you and can get about this business of
understanding what's behind this irrational feeling.
And when that's clear, the answers are obvious and I
don't need advice.
Irrational feelings make sense when we understand
what's behind them.
Perhaps that's why prayer works, sometimes, for some
people — because God is mute, and he/she doesn't
give advice or try to fix things. "They" just listen and
let you work it out for yourself.
So listen and just hear me. And, if you want to talk, wait
a minute for your turn, and I'll listen to you.

Contributors: Paul Radcliffe and Kathy Hunt

The Little Boy and the Old Man
Shel Silverstein

Said the little boy, "Sometimes I drop my spoon."
Said the little old man, "I do that too."
The little boy whispered, "I wet my pants."
"I do that too," laughed the little old man.
Said the little boy, "I often cry."
The old man nodded. "So do I."
"But worst of all," said the boy, "It seems
Grownups don't pay attention to me."
And he felt the warmth of a wrinkled old hand.
"I know what you mean," said the little old man.

A Light in the Attic

Contributor: Susan Schaefer

Love and Skill
John Ruskin

When love and skill work together expect a masterpiece.

Contributor: Jim Schoel

Morning Has Broken
Author Unknown

Morning has broken like the first morning.
Blackbird has spoken like the first bird.
Praise for the singing, praise for the morning.
Praise for them springing fresh from the word.

Sweet the rain's new fall, sunlit from heaven,
Like the first dew fall on the first grass.
Praise for the sweetness of the wet garden
Sprung in completeness where His feet pass.

Mine is the sunlight, mine is the morning,
Born of the one light Eden saw play.
Praise with elation, praise every morning:
God's recreation of the new day.

Old Universalist Hymn

Contributor: Jim Schoel

The Most Human Thing
Clarence Darrow

The most human thing we can do is comfort the afflicted and afflict the comfortable.

Contributor: Mike Stratton

The Mountains Can Be Reached
William O. Douglas

When man knows how to live dangerously, he is not afraid to die. When he is not afraid to die, he is, strangely, free to live. When he is free to live, he can become bold, courageous, self-reliant. There are many ways to learn how to live dangerously. Men of the plains have had the experience in the trackless blizzards that sweep in from the north. Those who go out in boats from Gloucester have known it in another form. The mountains that traverse this country offer still another way, and one that for many is the most exciting of all. The mountains can be reached in all seasons. They offer a fighting challenge to heart, soul and mind, both in summer and winter. If, throughout time, the youth of the nation accept the challenge the mountains offer, they will help keep alive in our people the spirit of adventure. That spirit is a measure of the vitality of both nations and men. A people who climb the ridges and sleep under the stars in high mountain meadows, who enter the forest and scale the peaks, who explore glaciers and walk ridges buried deep in snow — these people give their country some of the indomitable spirit of the mountains.

Of Men and Mountains

Contributor: Mike Stratton

Native Son
James Baldwin

It began to seem that one would have to hold in the mind forever two ideas which seemed to be in opposition. The first idea was acceptance, the acceptance, totally without rancor, of life as it is, and men as they are: in the light of this idea, it goes without saying that injustice is a commonplace. But this did not mean that one could be complacent, for the second idea was of equal power: that one must never, in one's own life, accept these injustices as commonplace but must fight them with all one's strength. This fight begins, however, in the heart and it now had been laid to my charge to keep my own heart free of hatred and despair.

Notes of a Native Son

Contributor: Jim Schoel

Never Act a Certain Way
Christopher Swift

Never act a certain way to fit in, for when all your acting is done, you will have forgotten who you really are.

Contributor: Bill Frankel

People Are Meant to be Loved
Author Unknown

People are meant to be loved.
Things are meant to be used.
Many of our problems stem from a disturbance of this
    reality.
People are used and things are loved.

Contributor: Nicole Richon-Schoel

The People in One's Life
Author Unknown

The people in one's life are like the pillars on one's porch you see life through. And sometimes they hold you up. And sometimes they lean on you, and sometimes it's just enough to know they're standing by.

Contributor: Jim Schoel

The Poor Scholar's Soliloquy
Dr. Stephen Corey

No, I'm not very good in school. This is my second year in the seventh grade and I'm bigger and taller than the other kids. They like me all right, though, even if I don't say much in the schoolroom, because outside I can tell them how to do a lot of things. They tag me around and that sort of makes up for what goes on in school.

I don't know why the teachers don't like me. They never have very much. Seems like they don't think you know anything unless you can name the book it comes out of. I've got a lot of books in my room at home — books like Popular Science Mechanical Encyclopedia, and the Sears' and Ward's catalogs, but I don't very often just sit down and read them through like they make us do in school. I use my books when I want to find something out, like whenever Mom buys anything secondhand I look it up in Sears' or Ward's first and tell her if she's getting stung or not. I can use the index in a hurry to find the things I want.

I guess I just can't remember names in history. Anyway, this year I've been trying to learn about trucks because my uncle owns three and he says I can drive one when I'm sixteen. I already know the horsepower and number of forward and backward speeds of twenty-six American trucks, some of them diesels, and I can spot each make a long way off. It's funny how that diesel works. I started to tell my teacher about it last Wednesday in science class when the pump we were using to make a vacuum in a bell jar got hot, but she said she didn't see what a diesel engine had to do with our experiment on air pressure, so I just kept still. The kids seemed interested, though. I took four of them around to my uncle's garage after school and we saw the mechanic, Gus, tearing a big truck diesel down. Boy, does he know his stuff!

I'm not very good in geography either. They call it economic geography this year. We've been studying the exports and imports of Chile all week but I couldn't tell you what they are. Maybe the reason is I had to miss school yesterday because my uncle took me and his big trailer truck down state about 200 miles and we brought almost 10 tons of stock to the Chicago market.

He had told me where we were going, and I had to figure out the highways to take and also the mileage. He didn't do anything but drive and turn where I told him to. Was that fun! I sat with a map in my lap and told him to turn south or southeast or some other direction. We made seven stops and drove over five hundred miles round trip. I'm figuring out what his oil cost and also the wear and tear on the truck — he calls it depreciation — so we'll know how much we made.

I even write out all the bills and send letters to the farmers about what their pigs and beef cattle brought at the stockyards. I only made three mistakes in 17 letters last time, my aunt said — all commas. She's been through high school and reads them over. I wish I could write school themes that way. The last one I had to write was on "What a Daffodil Thinks of Spring," and I just couldn't get going.

I don't do very well in school arithmetic, either. Seems I just can't keep my mind on the problems. We had one the other day like this: "If a 57-foot telephone pole falls across a cement highway so that 17-3/6 feet extend from one side and 14-6/17 feet from the other, how wide is the highway?" That seemed to me like an awfully silly way to get the width of a highway. I didn't even try to answer it because it didn't say whether the pole had fallen straight across or not.

Civics is hard for me, too. I've been staying after school trying to learn the "Articles of Confederation" for almost a week because the teacher said we couldn't be good citizens unless we did. I really tried, because I want to be a good citizen. I did hate to stay after school, though, because I and a bunch of us boys from the south end of town have been cleaning up the old lot across from Taylor's Machine Shop to make a playground out of it. We raised enough money collecting scrap this month to build a wire fence clear around the lot.

Dad says I can quit school when I am fifteen, and I am sort of anxious to because there are a lot of things I want to learn how to do and, as my uncle says, I'm not getting any younger.

Childhood Education

Contributor: Sally Jepson

Note: This classic first appeared in Childhood Education magazine in January of 1944. The late Dr. Stephen Corey was Dean of Faculty at Columbia University Teacher's College.

Popcorn
Andre Auew

"You can't get any popcorn, child. The machine is out of order. See, here is a sign on the machine."

But he didn't understand. After all, he had the desire, and he had the money, and he could see the popcorn in the machine. And yet somehow, somewhere, something was wrong because he couldn't get the popcorn.

The boy walked back with his mother, and he wanted to cry.

And Lord, I too felt like weeping, weeping for people who have become locked-in, jammed; broken machines filled with goodness that other people need, and want, and yet never come to enjoy, because somehow, somewhere, something had gone wrong inside.

Contributor: Jim Schoel

Popular
Garrison Keillor

One word I liked was popular. It sounded good, it felt good to say, it made lights come on in my mouth. I drew a rebus: a bottle of Nu-Grape + U + a Lazy Ike. Pop-u-lure. It didn't occur in our reading book, where little children did the right thing although their friends scoffed at them and where despised animals wandered alone and redeemed themselves through pure goodness and eventually triumphed to become Top Dog, the Duck of Ducks, the Grand Turtlissimo, the Greatest Pig Of Them All, which, though thrilling, didn't appeal to me so much as plain popular. "The popular boy came out the door and everybody smiled and laughed. They were glad to see him. They all crowded around him to see what he wanted to do."

Morning and afternoon, school recessed and we took to the playground; everyone burst out the door except me. Mrs. Meiers said, "Don't run! Walk!" I always walked. I was in no hurry, I knew what was out there. The girls played in front. Little girls played tag and stoop-ball, hopscotch, skipped rope; big girls sat under the pine tree and whispered. Some girls went to the swings. Boys went out back and played baseball, except for some odd boys who lay around in the shade and fooled with jackknives and talked dirty. I could go in the shade or stand by the backstop and wait to be chosen. Daryl and David always chose up sides and always chose the same people first, the popular ones. "Let somebody else be captain!" Jim said once. "How come you always get to choose?" They just smiled. They were captains, that was all there was to it. After the popular ones got picked, we stood in a bunch looking down at the dirt, waiting to see if our rating had changed. They took their sweet time choosing us, we had plenty of time to study our shoes. Mine were Keds, black, though white ones were more popular. Mother said black wouldn't show dirt. She didn't know how the wrong shoes could mark a person and raise questions in other people's minds. "Why do you wear black tennis shoes?" Daryl asked me once. He had me there. I didn't know. I guessed I was just that sort of person, whether I wanted to be or not. Maybe not showing dirt was not the real reason, the real reason was something else too terrible to know, which she would tell me someday. "I have something to tell you, son." She would say it. "No! No!" "Yes, I'm afraid it's true." "So that's why —" "Yes. I'm sorry I couldn't tell you before. I thought I should wait." "But can't I —" "No, I'm afraid not. We just have to make the best of it."

Nine boys to a side, four already chosen, ten positions left, and the captains look us over. They chose the popular ones fast ("Brian!" "Bill!" "Duke!" "John!" "Bob!" "Paul!" "Jim!" "Lance!"), and now the choice is hard because we're all so much the same: not so hot — and then they are down to their last grudging choices, slow kid for catcher and someone to stick out in right field where nobody hits it, except maybe two guys, and when they come to bat the captain sends the poor right-fielder to left, a long ignomini-ous walk. They choose the last ones two at a time, "You and you," because it makes no difference, and the remaining kids, the scrubs, the excess, they deal for as handicaps ("If I take him, then you gotta take him.") Sometimes I go as high as sixth, usually lower. Just once I'd like Daryl to pick me first. "Him! I want him! The skinny kid with the glasses and the black shoes! You! Come on!" But I've never been chosen with any enthusiasm.

Lake Wobegon Days

Contributor: Conrad Willeman

The Real Delinquents
Author Unknown

We read in the papers, we hear on the air
Of killing and stealing, of crime everywhere.
We sigh and we say as we notice the trend,
"This young generation, when will it end?"
But can we be sure that it's their fault alone,
That maybe most of it isn't really our own?
Too much money to spend, too much idle time,
Too many movies of passion and crime,
Too many books not fit to be read,

Too much of evil in what they hear said,
Too many children encouraged to roam,
By too many parents who won't stay home.
Kids don't make the movies, they don't write the books
    that paint a gay picture of gangsters and crooks.
They don't make the liquor, they don't run the bars, they
    don't pass the laws nor make high-speed cars.
They don't make the drugs that addle the brain.
It's all done by older folks greedy for gain.
Thus, in so many cases it must be confessed,
The label delinquent fits older folks best.

Found on Al's refrigerator

Contributor: Al Katz

A Relation with the Earth
Willa Cather

He had missed the deepest of all companionships, a
relation with the earth itself, with a countryside and a
people. That relationship, he knew, could not be gone after
and found; it must be long and deliberate, unconscious. It
must, indeed, be a way of living...and he had begun to
believe it the most satisfying tie men can have.

Lucy Gayheart

Contributor: Todd Tinkham

The Sea is Impersonal
Peter Osborne Willauer

The sea is impersonal, implacable, malign, and extraordi-
narily beautiful. One must dare greatly without foolhardi-
ness, and be at one's full powers to enjoy its best rewards —
and this can be done in a boat. There are no rewards or
punishments — no soul either good or bad — only consequences. You are responsible for your own action or inaction. The sea does not like you or dislike you or have any
human qualities — you deal with it by luck and skill and the
game really is to do what you have to do and stay out of the
way of the forces of nature that can wipe you out without
making the slightest difference one way or the other to
themselves.

Think ahead and be prepared for the unexpected. A good
seaman always plans ahead. Safety is an attitude. Everything
we do here in Outward Bound requires an awareness of

safety, and this awareness requires the right attitude of mind. No rules may be written if people do not think. Second thoughts often lead to better judgement.

HIOBS Instructor's Manual

Contributor: Peter Willauer

The Sky is Everlasting
Lao Tzu

> The sky is everlasting
> And the Earth is very old.
> Why so? Because the world
> Exists not for itself;
> It can and will live on.
> The wise man chooses to be last
> And so becomes first of all;
> Denying self, he too is saved.
> For does he not fulfillment find
> In being an unselfish man?

Contributor: Bruce Kezlarian

Solitude as an Art
Anne Morrow Lindbergh

It is a difficult lesson to learn today — to leave one's friends and family and deliberately practice the art of solitude for an hour or a day or a week... And yet, once it is done, I find there is a quality to being alone that is incredibly precious. Life rushes back into the void, richer, more vivid, fuller than before...

It is not physical solitude that actually separates one from other men, not physical isolation, but spiritual isolation. It is not the desert island nor the stony wilderness that cuts you from the people you love. It is the wilderness in the mind, the desert wastes in the heart through which one wanders lost and a stranger. When one is a stranger to oneself then one is estranged from others too. If one is out of touch with oneself, then one cannot touch others. How often, in a large city, shaking hands with my friends, I have felt the wilderness stretching between us. Both of us were wandering in arid wastes, having lost the springs that nourished us — or having found them dry. Only when one is connected to one's own core is one connected to others, I am beginning to discover. And for me, the core, the inner spring, can best be refound through solitude.

Gift from the Sea

Contributor: Jim Schoel

Specialization is For Insects
Robert A. Heinlein

A human being should be able to change a diaper, plan an invasion, butcher a hog, conn a ship, design a building, write a sonnet, balance accounts, build a wall, set a bone, comfort the dying, take orders, give orders, cooperate, act alone, solve equations, analyze a new problem, pitch manure, program a computer, cook a tasty meal, fight efficiently, die gallantly. Specialization is for insects.

The Notebooks of Lazarus Long

Contributor: Mike Stratton

Txamwork is Important
Author Unknown

Xvxn though my typxwritxr is an old modxl, it works quitx wxll, xxcxpt for onx of thx kxys. I wishxd many timxs that it workxd pxrfxctly. It is trux that thxrx arx forty-onx kxys that function wxll xnough, but just onx kxy not working makxs thx diffxrxncx. Somxtimxs it sxxms to mx that our organization is somxwhat likx this typxwritxr…not all thx pxoplx arx working. You say to yoursxlf wxll, I am only onx pxrson…I won't makx or brxak a program. But it doxs makx a diffxrxncx, bxcausx any program, to bx xffxctivx, nxxds thx activx participation of xvxry mxmbxr. So thx nxxt timx you think you arx only onx pxrson, and that your xfforts arx not nxxdxd, rxmxmmber my typxwritxr and thx story it txlls.

Contributor: Pat Shxcklxr

W. S. Coffin's Remarks on Service
W. S. Coffin

We like to think that good people will wake to their responsibilities and that Scrooge will finally sit down at the table with Tiny Tim. But greed and power are far more hard-hearted than even Dickens realized. And change in this world comes much more through the struggle and organiza-tion of the poor than it does from the belated decency of the rich.

HIOBS Newsletter, Fall 1988

Contributor: Jim Schoel

We Need Better Government
Wendell Berry

We need better government, no doubt about it. But we also need better minds, better friendships, better marriages, better communities. We need persons and households that do not have to wait upon organizations, but can make necessary changes in themselves, on their own.

A Continuous Harmony: Essays Cultural & Agricultural

Contributor: Todd Tinkham

What is a Friend?
Anonymous

What is a friend? I'll tell you. It is a person with whom you dare to be yourself. He seems to not want you to be better or worse. When you are with him, you feel as a prisoner feels who has been declared innocent. You do not have to be on your guard. You can say what you think; express what you feel. He is shocked at nothing, offended at nothing so long as it is genuinely you. He understands the little contradictions in your nature that lead others to misjudge you. With him, you breathe free. You can take off your coat, slip off your shoes. You can avow your vanities and envies and hates and vicious sparks, your meanness and absurdities, and in opening them up to him, they are lost, dissolved in the white ocean of his loyalty. He understands. You do not have to be careful. You can abuse him, neglect him, berate him. Best of all, you can keep still with him. He is like water that cleanses all you say. He is like wine that warms you to the bone. He understands. You can walk with him, sin with him, pray with him. Through and beneath it all he sees, knows, and loves you. A friend, I repeat, is the one with whom you dare to be yourself.

Contributor: Mike Stratton

The Worst Sin
George Bernard Shaw

The worst sin toward our fellow creatures is not to hate them, but to be indifferent to them; that's the essence of inhumanity.

Contributor: Jim Schoel

*H*UMOR

Dirt
Robert Service

Dirt is just matter out of place,
        So scientists aver;
But when I see a miner's face
        I wonder if they err.
For grit and grime and grease may be
        In God's constructive plan,
A symbol of nobility,
        The measure of a man.

There's nought so clean as honest dirt,
        So of its worth I sing;
I value more an oily shirt
        Than garmet of a king.
There's nought so proud as honest sweat,
        And though its stink we cuss,
We kid-glove chaps are in the debt
        Of those who sweat for us.

It's dirt and sweat that makes us folks
        Proud as we are today;
We owe our wealth to weary blokes
        Befouled by soot and clay.
And where you see a belly fat
        A dozen more are lean…
By God! I'd sooner doff my hat
        To washer-wife than queen.

So here's a song to dirt and sweat.
        A grace to grit and grime;
A hail to workers who beget
        The wonders of our time.

And as they gaze, though gutter-girt,
    To palaces enskied,
Let them believe, by sweat and dirt,
    They, too, are glorified.

Contributor: Joanne Maynard

Joanne's comments: Inevitably, after a day or two on the trail, complaints emerge from campers about their present state of filth. I've used this poem as a lighthearted way of saying, "Don't worry about it, enjoy!" It can spur some pretty funny conversation. The "dirtiest I've ever been" stories or "what is really under the bed at home," can make the group forget all about their hair and nails.

On a more serious level, the poem introduces a discussion about people in impoverished living conditions. Questions to ask are: Can we help? Should we help? What can we do? How do we keep from imposing ourselves and our values on others? What do people in other parts of the world endure that we feel we couldn't?

The Sloth

Theodore Roethke

In moving-slow he has no Peer.
You ask him something in his Ear,
He thinks about it for a Year;

And, then, before he says a Word
There, upside down (unlike a Bird),
He will assume that you have Heard —

A most Ex-as-per-at-ing Lug.
But should you call his manner Smug,
He'll sigh and give his Branch a Hug;

Then off again to Sleep he goes,
Still swaying gently by his Toes,
And you just *know* he knows he knows.

Contributor: Joanne Maynard

Joanne's comments: Have you ever been with a group that took double the expected time to do everything? Have you ever felt that as a leader you were stuck in low gear and couldn't get out? This poem can be used on those occasions.

Be careful not to use it to ridicule, but rather as a light way of taking a look at behavior. I've used it with kids as well as adults. If it is presented in the right spirit, it can be a fun way to giggle at ourselves.

Questions to ask: What are the advantages of taking

*your time? What are the disadvantages? Did you miss
anything because you were slow? Did you gain anything?
Is it important to be able to laugh at ourselves? Why?
Does that mean we don't take ourselves seriously if we do
laugh at ourselves?'*

*If you have a group that needs and wants to move
slowly you could try having them compile lists of individual observations and then come together to share
them. Try just a list of smells or sounds. It is fascinating
to discover the different things that people tune into.*

A Blessing
Author Unknown

> May neither drought nor rain nor blizzard
> Disturb the Joy-juice in your gizzard!
> And may you camp where wind won't hit you,
> Where snakes won't bite and bears won't git you.

Contributor: Jim Schoel

The Cremation of Sam McGee
Robert Service

> *There are strange things done in the midnight sun*
>   *By the men who moil for gold;*
> *The Arctic trails have their secret tales*
>   *That would make your blood run cold;*
> *The Northern Lights have seen queer sights,*
>   *But the queerest they ever did see*
> *Was that night on the marge of Lake Lebarge*
>   *I cremated Sam McGee.*

> Now Sam McGee was from Tennessee, where the cotton
>   blooms and blows.
> Why he left his home in the South to roam 'round the
>   Pole, God only knows.
> He was always cold, but the land of gold seemed to hold
>   him like a spell;
> Though he'd often say in his homely way that "he'd
>   sooner live in hell."

> On a Christmas Day we were mushing our way over the
>   Dawson trail.
> Talk of the cold! through the parka's fold it stabbed like a
>   driven nail.
> If our eyes we'd close, then the lashes froze till sometimes
>   we couldn't see;
> It wasn't much fun, but the only one to whimper was Sam
>   McGee.

And that very night, as we lay packed tight in our robes
    beneath the snow,
And the dogs were fed, and the stars o'erhead were
    dancing heel and toe,
He turned to me, and "Cap," says he, "I'll cash in this
    trip, I guess;
And if I do, I'm asking that you won't refuse my last
    request."

Well, he seemed so low that I couldn't say no; then he
    says with a sort of moan:
"It's the cursed cold, and it's got right hold till I'm chilled
    clean through to the bone.
Yet 'taint being dead — it's my awful dread of the icy
    grave that pains;
So I want you to swear that, foul or fair, you'll cremate
    my last remains."

A pal's last need is a thing to heed, so I swore I would not
    fail;
And we started on at the streak of dawn; but God! he
    looked ghastly pale.
He crouched on the sleigh, and he raved all day of his
    home in Tennessee;
And before nightfall a corpse was all that was left of Sam
    McGee.

There wasn't a breath in that land of death, and I hurried,
    horror-driven,
With a corpse half hid that I couldn't get rid, because of a
    promise given;
It was lashed to the sleigh, and it seemed to say: "You
    may tax your brawn and brains,
But you promised true, and it's up to you to cremate
    those last remains."

Now a promise made is a debt unpaid, and the trail has
    its own stern code.
In the days to come, though my lips were dumb, in my
    heart how I cursed that load.
In the long, long night, by the lone firelight, while the
    huskies, round in a ring,
Howled out their woes to the homeless snows — O God!
    how I loathed the thing.

And every day that quiet clay seemed to heavy and
    heavier grow;
And on I went, though the dogs were spent and the grub
    was getting low;

The trail was bad, and I felt half mad, but I swore I would not give in;
And I'd often sing to the hateful thing, and it hearkened with a grin.

Till I came to the marge of Lake Lebarge, and a derelict there lay;
It was jammed in the ice, but I saw in a trice it was called the "Alice May."
And I looked at it, and I thought a bit, and I looked at my frozen chum;
Then "Here," said I, with a sudden cry, "is my cre-ma-tor-eum."

Some planks I tore from the cabin floor, and I lit the boiler fire;
Some coal I found that was lying around, and I heaped the fuel higher;
The flames just soared, and the furnace roared — such a blaze you seldom see;
And I burrowed a hole in the glowing coal, and I stuffed in Sam McGee.

Then I made a hike, for I didn't like to hear him sizzle so;
And the heavens scowled, and the huskies howled, and the wind began to blow.
It was icy cold, but the hot sweat rolled down my cheeks, and I don't know why;
And the greasy smoke in an inky cloak went streaking down the sky.

I do not know how long in the snow I wrestled with grisly fear;
But the stars came out and they danced about ere again I ventured near;
I was sick with dread, but I bravely said; "I'll just take a peep inside.
I guess he's cooked, and it's time I looked,"... then the door I opened wide.

And there sat Sam, looking cool and calm, in the heart of the furnace roar;
And he wore a smile you could see a mile, and he said: "Please close that door.
It's fine in here, but I greatly fear you'll let in the cold and storm —
Since I left Plumtree, down in Tennessee, it's the first time I've been warm."

> There are strange things done in the midnight sun
> > By the men who moil for gold;
> The Arctic trails have their secret tales
> > That would make your blood run cold;
> The Northern Lights have seen queer sights,
> > But the queerest they ever did see
> Was that night on the marge of Lake Lebarge
> > I cremated Sam McGee.

Contributor: Jim Schoel

Deo Gratias
John A. Galm

There's a German short-form
Grace for meals
That'll get you to heaven
With almost no more sweat:

> Fur Dies und das,
> Deo Gratias.

As I sit in my garden,
Saturday morning,
I'm full of thanks
That God isn't a body-builder
Or an exercise freak.
After six days of madcap creating
He was pooped. And rested.

And then the Jews and Catholics couldn't get together
   about which was the
Seventh Day —
Giving us a
Two-day weekend.

Keen.
Deo Gratias.

Lower Stumpf Lake Review

Contributor: Lisa Galm

The Dry Ones
Rufus Little and Karl Rohnke

The only people who mind getting wet are the ones who are dry.

Contributor: Karl Rohnke

Hug O' War

Shel Silverstein

> I will not play at tug o' war
> I'd rather play at hug o' war
> Where everyone hugs
> Instead of tugs
> Where everyone giggles
> And rolls on the rug,
> Where everyone kisses,
> And everyone grins,
> And everyone cuddles,
> And everyone wins.

A Light in the Attic

Contributor: Susan Schaefer

I Don't Want to Achieve

Woody Allen

I don't want to achieve immortality through my work. I want to achieve immortality through not dying.

Contributor: Karl Rohnke

If It Doesn't Kill You

Karl Rohnke

> If it doesn't kill you it'll make you strong.

A paraphrase of Nietzsche

Contributor: Karl Rohnke

If It's Worth Doing

Karl Rohnke

> If it's worth doing, it's worth overdoing.

Contributor: Karl Rohnke

If You Are Unhappy

Author Unknown

Once upon a time, there was a nonconforming sparrow who decided not to fly south for the winter. However, soon the weather turned so cold that he reluctantly started to fly south. In a short time ice began to form on his wings and he fell to earth in a barnyard, almost frozen. A cow passed by and crapped on the little sparrow. The sparrow thought it was the end. But, the manure warmed him and defrosted his wings. Warm and happy, able to breathe, he started to sing.

Just then, a large cat came by and hearing the chirping, investigated the sounds. The cat cleared away the manure, found the chirping bird and promptly ate him.

The moral of the story:

1. Everyone who shits on you is not necessarily your enemy.

2. Everyone who gets you out of shit is not necessarily your friend.

3. And, if you're warm and happy in a pile of shit, keep your mouth shut.

Contributor: Jim Schoel

In Humor There is Truth
Ralph Nader

In humor, there is truth. We need to take humor more seriously.

Contributor: Karl Rohnke

The Insecure Camper
A Bounder

For one thing, I forgot my sleeping bag. This hike is too steep and I'm thirsty. That dumb kid in front of me keeps tripping me up, and these boots are making hamburger out of my feet. When I got to the top of Mt. Moosilauke, I froze, because I dropped my sweater in the brook. Everyone says, "You'll laugh about this in five years," but I think that's a truckload of bull.

If you think that's bad, when we did Spree, I got stuck with two jerks in my Bounder's Tent, and I hate frog's legs. It's boiling hot, and we can't swim in the lake and Gypsy moths make me sick. I guess I just had an off day.

Contributor: Mike Stratton

Last Words
Henry David Thoreau

"Moose. Indians." (Before this Thoreau had been asked if he had made his peace with God and replied, "I was not aware that we had ever quarreled.")

Thoreau's last words before his death

Contributor: Tom Zierk

Life is Like a Fan

Author Unknown

Life is like a fan: if you are up front, it's a breeze. If you're in back, it sucks.

Contributor: Mike Stratton

Love in a Tide Pool

Jane van Alst

There are many ways to do it,
Think of flowers, birds and bees,
But if you really want an expert
Let me take you to the seas.
Oh, the algae look like simple things,
Adrifting in the bays,
But when it comes to reproduction
Why they've thought of all the ways.
The lower types form akinetes,
A stand-offish way to be
While their higher swinging cousins
Practice good, old isogamy.
Let's slip into something gelatinous,
The aplanogametes say,
And have a little syngamy,
Before we float away.
Now isogamy's not bad at all
As a method used for fusing,
But most find heterogamy
A damn sight more amusing.
At an intertidal orgy
You can hear the Fucus gloat
While they pull out all paraphyses
And make a new zygote.
Now when the sun is shining
And the water's bright as day
Oogamy's the answer
For the sperm can find their way.
But when the sun sets slowly
And the ocean depths get dim
Then they hope that they're monoecious
'Cause who can see to swim?
So don't ever snub a seaweed,
Or give a kelp the hex,
'Cause man, like they invented it
And we just named it, SEX!

Contributor: Susan Schaefer

My Heart Wants Roots

E.Y. Harburg

> My heart wants roots.
> My mind wants wings.
> I cannot bear
> Their bickerings.

Rhymes for the Irreverent

Contributor: Bonnie Hannable

Never

Shel Silverstein

> I've never roped a Brahma bull,
> I've never fought a duel,
> I've never crossed the desert
> On a lop-eared, swayback mule,
> I've never climbed an idol's nose
> To steal a cursed jewel.
> I've never gone down with my ship
> Into the bubblin' brine,
> I've never saved a lion's life
> And then had him save mine,
> Or screamed Ahoooo while swingin' through
> The jungle on a vine.
> I've never dealt draw poker
> In a rowdy lumber camp,
> Or got up at the count of nine
> To beat the world's champ,
> I've never had my picture on
> A six-cent postage stamp.
> I've never scored a touchdown
> On a ninety-nine yard run,
> I've never winged six Daltons
> With my dying brother's gun...
> Or kissed Miz Jane, and rode my hoss
> Into the setting sun.
> Sometimes I get so depressed
> 'Bout what I haven't done.

A Light in the Attic

Contributor: Susan Schaefer

Regarding Marlene Dietrich

Ernest Hemingway

> Never confuse motion with action.

Contributor: Jim Schoel

Return to the Womb

Woody Allen

I have an intense desire to return to the womb. Anybody's.

Contributor: Todd Tinkham

The Saga of Baba Fats

Author Unknown

There once was a boy called Gimmesome Roy. He was
   nothing like me or you.
Cause laying back and getting high was all he cared to do.
As a kid, he sat down in his cellar, sniffing airplane glue.
And then he smoked bananas, which was then the thing
   to do.
He tried aspirin and Coca-Cola, breathed helium on the
   sly,
And his life was just one endless search to find that
   perfect high.
But grass just made him want to lay back and eat choco-
   late-chip pizza all night,
And the great things he wrote while he was stoned looked
   like shit in the morning light.
And speed just made him rap all day, reds just laid him
   back,
And Cocaine Rose was sweet to his nose, but her price
   nearly broke his back.
He tried PCP and THC but they didn't quite do the trick,
And poppers nearly blew his heart and mushrooms made
   him sick
Acid made him see the light, but he never remembered
   long.
And hashish was just a little too weak, and smack was a
   lot too strong.
And Quaaludes made him stumble, and booze just made
   him cry,
Till he heard of a cat named Baba Fats who knew of the
   perfect high.
Now Baba Fats was a hermit cat who lived up in Nepal,
High on a craggy mountaintop, up a sheer icy wall.
"But hell," says Roy, "I'm a healthy boy, and I'll crawl or
   climb or fly,
But I'll find that guru who'll give me the clue as to what's
   the perfect high."
So out and off goes Gimmesome Roy to the land that
   knows no time.
Up a trail no man could conquer to a cliff no man could
   climb.

For fourteen years he tries that cliff, then back down
    again he slides,

Then sits — and cries — and climbs again, pursuing that
    perfect high.

He's grinding his teeth, he's coughing blood,

he's aching and shaking and weak,

As starving and sore and bleeding and tore he reaches the
    mountain peak.

And his eyes blink red like a snow-blind wolf and he
    snarls the snarl of a rat,

As there in perfect repose and wearing no clothes — sits
    the godlike Baba Fats.

"What's happening, Fats?" says Roy with joy, "I come to
    state my biz.

I hear you're hip to the perfect trip. Please tell me what it
    is.

For you can see," says Roy to he, "that I'm about to die.

So, for my last ride, Fats, how can I achieve that perfect
    high?"

"Well, dog my cats," says Baba Fats, "here's one more
    burnt-out soul,

Who's looking for some alchemist to turn his trip to gold.

But you won't find it in no dealer's stash or on the
    druggist's shelf

Son, if you seek the perfect high — find it in yourself."

"Why you (blankety blank)" screamed Gimmesome Roy,
    "I've climbed through rain and sleet.

I've lost three fingers off my hands and four toes off my
    feet.

I've braved the lair of the polar bear and tasted the
    maggot's kiss.

Now you tell me the high is in myself, what kind of shit
    is this?

Look! My butt's froze off," says Roy, "and I've heard all
    kinds of crap

But I didn't dream in fourteen years to listen to that
    sophomore rap.

And I didn't crawl up here to hear that the high is on the
    natch.

So you tell me where the real stuff is or I'll kill you guru
    ass."

"OK, OK," says Baba Fats, "you're forcing it out of me.

There is a land beyond the sun that's known as Zaboli.

A wretched land of stone and sand where snakes and
    lizards scree

Where the devil's guardian guards the mystic Tzu Tzu
    tree.

Once every year it blooms, the flower as white as the Key
    West sky.

And he who eats of this Tzu Tzu flower will know the
   perfect high,
For the rush comes on like a tidal wave and hits like the
   blazing sun,
And the high, it lasts a lifetime and the down don't ever
   come.
But the Zaboli land is ruled by a giant who stands twelve
   cubits high.
With eyes of red in its hundred heads, he waits for the
   passers-by.
And you must slay the red-eyed giant and then swim the
   River of Slime
Where the beasts, they wait to feast on those who journey
   by.
And if you survive the giant and beasts and swim the
   slimy sea,
There's a blood-drinking witch who sharpens her teeth as
   she guards the Tzu Tzu tree."
"To hell with your witches and giants," laughs Roy, "To
   hell with the beasts of the sea.
As long as the Tzu Tzu flower blooms, some hope still
   blooms for me."
And with tears of joy in his snow-blind eye, Roy hands
   the guru a five,
Then back down the icy mountain he crawls, pursuing
   that perfect high.
"Well, that is that," says Baba Fats, sitting back down on
   his stone,
Facing another thousand years of talking to God alone.
"It seems, Lord," says Fats, "it's all the same, old men or
   bright-eyed youth.
It's always easier to sell them some shit than it is to give
   them the truth."

Contributor: Scott Garman

Whatif

Shel Silverstein

Last night, while I lay thinking here, Some Whatifs
   crawled inside my ear
And pranced and partied all night long
And sang their same old Whatif song:
Whatif I'm dumb in school?
Whatif they've closed the swimming pool?
Whatif I get beat up?
Whatif there's poison in my cup?
Whatif I start to cry?
Whatif I get sick and die?
Whatif I flunk that test?

Whatif green hair grows on my chest?
Whatif nobody likes me?
Whatif a bolt of lightning strikes me?
Whatif I don't grow taller?
Whatif my head starts getting smaller?
Whatif the fish won't bite?
Whatif the wind tears up my kite?
Whatif they start a war?
Whatif my parents get divorced?
Whatif the bus is late?
Whatif my teeth don't grow in straight?
Whatif I tear my pants?
Whatif I never learn to dance?
Everything seems swell, and then
The nighttime Whatifs strike again!

A Light in the Attic

Contributor: Susan Schaefer

Wood Song (à la Goodspeed and friends)
Author Unknown

In the fall of the year when the winter's drawing near, and the days are clear, it certainly isn't good to sit by the fire and watch yourself get higher when you should be cutting more wood. From November to March when the winter winds are harsh and the fields and the marsh are covered up with snow, you drudge to shed and have to scratch your head 'cause the dad-blamed pile's gettin' low.

On wood (on wood), dry wood (dry wood), there ain't a stove in the world gonna do you any good. Without wood (hard wood), we could (you should), be out cuttin' more wood.

Now the kindling is dwindling, the bottom log is soggy and the wicks 'n sticks and racks and stacks make you wonder where they go. Barnfulls and barnfulls that only last a week or so, and then you're hurtin' for wood. Now the Strindley and the Yodel brand are made so far across the sea while fisherline and timberline are made here in the country but of all the rest that've stood the test, the one I like the very best is the stove my uncle Wade made for me.

He took an oil drum and welded some piping from the septic tank. Fore and aft he cut a draft then he built a damper crank of an old broom in the back room and painted it fire engine red and said, "Now watch it consume: Your wood (hard wood), dry wood (stove wood), there ain't a stove in the world gonna do ya' any good without wood (stove wood), we could (you should), be out cuttin' more wood."

Now the old timers say to split a little every day and stack it away to season it well but from March to November I rarely do remember and December will find me in a rut. When spring rolls around and I spade the muddy ground I have often found I laid my saw away. The shed is empty and yet you may make the bet I'll forget to be cuttin' any wood.

Said wood (hard wood), stove wood (stove wood), there ain't a stove in the world gonna do ya' any good without wood (hard wood), we could (you should) be out cuttin' more, throwin' it in the oil drum, whaddaya think the saw is for, you will always need some more WOOD.

Contributor: Mike Stratton

Wrinkles
The Talking Moose

If you don't have wrinkles, you haven't laughed enough.

Contributor: Conrad Willeman

Journey

Starting Out
George Back

There is something exciting in the first start, even upon an ordinary journey. The bustle of preparation — the act of departing, which seems like a decided step taken — the prospect of change, and consequent stretching out of the imagination — have at all times the effect of stirring the blood, and giving a quicker motion to the spirits. It may be conceived then with what sensations I set forth on my journey into the Arctic wilderness. I had escaped from the wretchedness of a dreary and disastrous winter — from scenes and tales of suffering and death — from wearisome inaction and monotony — from disappointment and heart-sickening care. Before me were novelty and enterprise; hope, curiosity, and the love of adventure were my companions; and even the prospect of difficulties and dangers to be encountered, with the responsibility inseparable from command, instead of damping rather heightened the enjoyment of the moment. In turning my back on the Fort, I felt my breast lightened, and my spirit, as it were, set free again; and with a quick step, Mr. King and I (for my companion seemed to share in the feeling) went on our way rejoicing.

Narrative of the Arctic Land Expedition, 1836

Contributor: Bob Henderson

Bob's comments: I have used this passage by George Back at the beginning of a group travel experience. Whether one is departing the Comforts (be they as they may) or wintering over at Fort Enterprise before the final push into the Arctic unknown, or a weekend trip in newly explorable local bush, the feeling of excitement in the act of departing are common. The passage is best read to an attentive group just before donning the snowshoe or boarding canoes. Ahhh...to be an active part of a great tradition.

Experiential Teacher's Paradox

Donald Schön

It's as though the teacher said something like this: "I can tell you that there's something you need to know and I can tell you that with my help you can probably learn it. But I cannot tell you what it is in a way that you can now understand. You must be willing therefore, to undergo certain experiences as I direct you to undergo them, so that you can learn what it is that you need to know and what I mean by the words that I use. Then and only then can you make an informed choice about whether you wish to learn this new competence. If you are unwilling to step into this new experience without knowing ahead of time what it will be like, then I cannot help you. You must trust me."

From an address at Queen's University

Contributor: James Raffan

tural school, whom he saw to have complicated notions of what it meant to "think architecturally" — what they wanted the students to do — but who were inept at describing what they meant by this lofty term. In many ways, this situation parallels that of an experiential educator who knows implicitly the power of a particular kind of experience, but who, lacking the necessary language or student readiness, is incapable of articulating exactly what the power is. Although it's imposing — "You must trust me" — I've found this reading useful when embarking on an experience whose benefits I know, but can't describe. It recognizes the synergistic nature of people and experience, but it also calls for and focuses participant commitment on the exercise. And, best of all, it heightens participants' feeling of accomplishment when, in a debriefing of the experience, they begin to "see" what the instructor meant.

Hand of Cards
Woodrow Wilson Sayre

Each one of us is dealt a hand of cards by life. It's not so much the hand you get dealt but what you do with what you've got.

Four Against Everest

Contributor: Phil Salzman

Phil's comments: I use this to introduce the idea of the uncontrollable elements in life, and then try to get beyond them to how you've got to play your hand, make decisions, and develop your own vision, power and journey. The individual owns this. That is what character development is about...that we're each empowered to make decisions that we have to own...not denying that some have better hands than others. That is the nature of life.

I Tramp a Perpetual Journey
Walt Whitman

I know I have the best of time and space, and was never
    measured, and never will be measured.
I tramp a perpetual journey — (come listen all!)
My signs are a rain-proof coat, good shoes, and a staff cut
    from the woods;
No friend of mine takes his ease in my chair;
I have no chair, no church, no philosophy;
I lead no man to a dinner-table, library, or exchange;
But each man and each woman of you I lead upon a
    knoll,

My left hand hooking you round the waist, My right hand
    pointing to landscapes of continents, and a plain
    public road.
Not I — not any one else, can travel that road for you,
You must travel it for yourself.
It is not far— it is within reach;
Perhaps you have been on it since you were born, and did
    not know;
Perhaps it is everywhere on water and on land.
Shoulder your duds, dear son, and I will mine, and let us
    hasten forth,
Wonderful cities and free nations we shall fetch as we go.
If you tire, give me both burdens, and rest the chuff of
    your hand on my hip,
And in due time you shall repay the same service to me;
For after we start, we never lie by again.
This day before dawn I ascended a hill, and look'd at the
    crowded heaven,
And I said to my Spirit, *When we become the enfolders of
    these orbs, and the pleasure and knowledge of every-
    thing in them, shall we be fill'd and satisfied then?*
And my Spirit said, *No, we but level that lift, to pass and
    continue beyond.*
You are also asking me questions, and I hear you;
I answer that I cannot answer — you must find out for
    yourself.
Sit a while, dear son;
Here are biscuits to eat, and here is milk to drink;
But as soon as you sleep, and renew yourself in sweet
    clothes, I kiss you with a good-bye kiss, and open the
    gate for your egress hence.
Long enough have you dream'd contemptible dreams;
Now I wash the gum from your eyes;
You must habit yourself to the dazzle of the light, and of
    every moment of your life.
Long have you timidly waded, holding a plank by the
    shore;
Now I will you to be a bold swimmer, To jump off in the
    midst of the sea, rise again, nod to me, shout, and
    laughingly dash with your hair.

Song of Myself (Stanza #46) Leaves of Grass

Contributor: Susan St. John-Rheault

*Susan's comments: Song of Myself is one of my
favorite readings. I have read it to young adults as they go
on solo, or as they head home, pointing out the magnifi-
cence of the journey, that it must be done alone and yet
that we all have partners to help and be helped by.*

Ithaka

C.P. Cavafy

When you set out for Ithaka
Ask that your way be long, Full of adventure, full of
   instruction.
The Laistrygonians and the Cyclops, angry Poseidon —
   do not fear them;
Such as these you will never find
As long as your thought is lofty, As long as a rare emotion
Touch your spirit and your body.
The Laistrygonians and the Cyclops,
Angry Poseidon — you will not meet them
Unless you carry them in your soul, Unless your soul
   raise them up before you.

Ask that your way be long
At many a summer dawn to enter-
With what gratitude, what joy!
Ports seen for the first time;
To stop at Phoenician trading centers, And to buy good
   merchandise:
Mother of pearl and coral, amber and ebony, And
   sensuous perfumes of every kind.
Buy as many sensuous perfumes as you can, Visit many
   Egyptian cities
To learn and learn from those who have knowledge.

Always keep Ithaka fixed in your mind
Your arrival there is what you are destined for.
But do not in the least hurry the journey.
Better that it last for years
So that when you reach the island you are old
Rich with all that you have gained on the way, Not
   expecting Ithaka to give you wealth.
Ithaka has given you the splendid voyage.
Without her you would never have set out, But she has
   nothing more to give you.

And if you find her poor, Ithaka has deceived you.
So wise have you become, of such experience, That
   already you will have understood
What these Ithakas mean.

Contributor: John K. Spencer

*John's comments: At 0830 we were still standing on
the Valley Cove float in a cold southeasterly drizzle, lead-
gray clouds scudding overhead. Both pulling boats were
loaded for final expedition, now two hours past the
planned departure time. Duffels and food cans were
carelessly packed and arranged; on top of them lay still-
unwashed breakfast pans. The watches were engaged in
an endless relay to and from the Boathouse for one*

forgotten or misplaced essential after another.

Forecasts consistently predicted rain and a cold wind for the next few days. No one wanted to leave; we knew our course would not return us to Hurricane again. Thinking of dry tents, fireplaces and Rick's cooking, everyone was swooning to the island Lorelei's song.

Finally, the last person stepped aboard, and my reading book fell open to the marked page. We asked that the trip be long, full of challenge. We dreamed of adventure together, not of the angry Poseidon. Rockland's riches would not be our goal or promise.

For the first time that morning, all eyes looked southeast together. A minute of silence was broken by a dockline splashing into the water. A captain called, "Toss oars!" And someone said, "Let's get outta here!"

Psalm 91

He who dwells in the shelter of the Most High
will rest in the shadow of the Almighty.
I will say of the Lord, He is my refuge and my fortress, my
    God, in whom I trust.
Surely, he will save you from the fowler's snare
and from the deadly pestilence.
He will cover you with his feathers, and under his wings
    you will find refuge;
his faithfulness will be your shield and rampart.
You will not fear the terror of night, nor the arrow that
    flies by day, nor the pestilence that stalks in the
    darkness, nor the plague that destroys at midday.
A thousand may fall at your side, ten thousand at your
    right hand, but it will not come near you.
You will only observe with your eyes
and see the punishment of the wicked.
If you make the Most High your dwelling—
even the Lord, who is my refuge—
then no harm will befall you, no disaster will come near
    your tent.
For he will command his angels concerning you
to guard you in all your ways;
they will lift you up in their hands, so that you will not
    strike your foot against a stone.
You will tread upon the lion and the cobra;
you will trample the great lion and the serpent.
"Because he loves me," says the Lord, "I will rescue him;
I will protect him, for he acknowledges my name.
He will call upon me, and I will answer him;
I will be with him in trouble, I will deliver him and honor
    him.

With long life will I satisfy him
and show him my salvation."

Contributor: Susan Schaefer

Susan's comments: I've found myself using this Psalm on almost every La Vida trip I've led. After seeing a group go through the new experiences of climbing rocks, carrying heavy packs, and hiking mountain trails, it always seems appropriate to send them off on their solo time with these thoughts. Participants often express fears of being alone, or fighting off bugs, or facing a long dark night. This reading helps remind them of the times during the trip they've felt God's love and protection. This Psalm refers to the escape of the Israelites from Egypt and journey into the wilderness, but I sometimes make it more relevant by changing a few key words: arrow=mosquito, pestilence=giardia, etc.

Tucker Foundation Credo

William Jewett Tucker

Be not content with the commonplace in character
anymore than with the commonplace in ambition or
intellectual attainment.
Do not expect that you will make any lasting or
very strong impression on the world through
intellectual power without the use of an equal
amount of conscience and heart.

Dartmouth College, wooden sign at entrance of college hall

Contributor: Mike Stratton

Mike's comments: I never went by this sign without reading and contemplating its wisdom and advice as I approached my June 14, 1969 graduation and important decisions about the draft, grad school, job-career search. The end result was going to work on June 15, 1969 at the Tucker Foundation's Dartmouth Outward Bound Center which led to over twenty years of committed involvement in experiential education!

Art as Act

Edmund Carpenter

When art becomes inseparable from daily living — the
way a woman prepares a meal, speaks to her children,
decorates her home, makes love, laughs — there is no "art,"
for all life is artistic.

They Became What They Beheld

Contributor: Mike Stratton

Being First
Dan Baker

Being first is not my style
I like to stop and feel awhile
Under the moon's thin-lipped smile
I find a peace that lasts for miles.
By myself I'm not alone
No TV guide, no telephone
The quiet noise in the trees
Cuts my bonds, puts my mind at ease.
You know my journey never ends
Long as I am I will be travelling
To know myself and know my friends
To heal some hurts and make amends
Down this rocky trail I find
The roughest path is in my mind
Down this rocky trail I find
The roughest path is in my mind.

Contributor: Bonnie Hannable

Birthday Thought
Eubie Blake

If I'd known how long I was going to live, I would've taken better care of myself.

Contributor: Bob Ryan

Burned Out?
Edward Abbey

Do not burn yourselves out. Be as I am. A reluctant enthusiast and part-time crusader. A half-hearted fanatic. Save the other half of yourselves for pleasure and adventure. It is not enough to fight for the west. It is even more important to enjoy it while you can, while it's still there. So get out there, hunt, fish, mess around with your friends, ramble out yonder and explore the forests, encounter the griz, climb a mountain, bag the peaks, run the rivers, breathe deep of that yet sweet and elusive air. Sit quietly for a while and contemplate the precious stillness of the lovely, mysterious, and awesome space. Enjoy yourselves. Keep your brain in your head and your head firmly attached to the body, the body active and alive. And I promise you this one sweet victory over our enemies, over those desk-bound people with their hearts in safe deposit boxes and their eyes hypnotized by their desk calculators. I promise you this: You will outlive the bastards.

A speech to environmentalists in Missoula, Montana, 1978

Contributor: Brian Pritchard

Call Me Ishmael
Herman Melville

Call me Ishmael. Some years ago — never mind how long precisely — having little or no money in my purse, and nothing particular to interest me on shore, I thought I would sail about a little and see the watery part of the world. It is a way I have of driving off the spleen, and regulating the circulation. Whenever I find myself growing grim about the mouth; whenever it is a damp, drizzly November in my soul; whenever I find myself involuntarily pausing before coffin warehouses, and bringing up the rear of every funeral I meet; and especially whenever my hypos get such an upper hand of me, that it requires a strong moral principle to prevent me from deliberately stepping into the street, and methodically knocking people's hats off — then, I account it high time to get to sea as soon as I can. This is my substitute for pistol and ball. With a philosophical flourish Cato throws himself upon his sword; I quietly take to the ship. There is nothing surprising in this. If they but knew it, almost all men in their degree, some time or other, cherish very nearly the same feelings towards the ocean with me.

Moby Dick

Contributor: Jim Schoel

Celebrate What is Constant
James Baldwin

It is the responsibility of free men to trust and to celebrate what is constant — birth, struggle, and death are constant — and so is love, though we may not always think so — and to apprehend the nature of change, to be able and willing to change. I speak of change not on the surface but in the depths — change in the sense of renewal. But renewal becomes impossible if one supposes things to be constant that are not — safety, for example, or money, or power. One clings then to chimeras, by which one can only be betrayed, and the entire hope — the entire possibility — of freedom disappears.

The Fire Next Time

Contributor: Jim Schoel

Climbing Is a Sport
Author Unknown

Climbing is a sport, but climbing in the mountains, like ocean racing or crossing a desert takes place in different conditions from those of common sports. A climb is not a sort of a game which can be stopped at any time. Even if you are at the limits of endurance, if your feet feel like lead, if nothing but extreme effort of will keeps you going, even if lightning is flashing across the sky, you cannot sit down and say, "I've had enough. I'm giving up. I quit." And even when you do get to the top, the rock is still not half finished. This is undoubtedly the hardest of rules to accept, but it is nevertheless an attraction: on every crest the climber must ride his whole self.

Contributor: Mike Stratton

The Climb Seems Endless
Dag Hammarskjöld

When the morning's freshness has been replaced by the weariness of mid-day, when the leg muscles quiver under the strain, the climb seems endless, and, suddenly, nothing will go quite as you wish — it is then that you must not hesitate.

Markings

Contributor: Mike Stratton

Conifers
Ernest Thompson Seton

The conifers illustrate better than others of our trees the process and plan of growth. Thus a seedling pine has a tassel or two at the top of a slender shoot, next year it has a second shoot and a whorl corresponding exactly with its vigor that season, until the tree is so tall that the lower whorls die, and their knots are overlaid by fresh layers of timber. The timber grows smoothly over them, but they are just the same, and anyone carefully splitting open one of these old forest patriarchs, can count on the spinal column the years of growth, and learn in a measure how it fared each season.

Forester's Manual

Contributor: Scott Garman

Each Second We Live

Pablo Casals

Each second we live in a new and unique moment of the Universe, a moment that never was before and will never be again. And what do we teach our children in school? We teach them that two and two make four, and that Paris is the capital of France. When will we also teach them what they are? We should say to each of them: do you know what you are? You are a marvel. You are unique. In all of the world there is no other child exactly like you. In the millions of years that have passed there has never been another child like you. And look at your body — what a wonder it is! Your legs, your arms, your cunning fingers, the way you move! You may become a Shakespeare, a Michelangelo, a Beethoven. You have the capacity for anything. Yes, you are a marvel. And when you grow up, can you then harm another who is, like you, a marvel? You must cherish one another. You must work — we must all work — to make this world worthy of its children.

Joys and Sorrows: reflections by Pablo Casals, as told to Albert E. Kahn

Contributor: Susan Schaefer

Ecclesiastes 1:4–7

King Solomon

Generations come and generations go, but the earth remains forever. The sun rises and the sun sets, and hurries back to where it rises. The wind blows to the south and turns to the north; round and round it goes, ever returning on its course. All streams flow into the sea, yet the sea is never full. To the place the streams come from, there they return again.

Bible, New International Version

Contributor: Jim Schoel

Education and Experience

Pete Seeger

Do you know the difference between education and experience? Education is when you read the fine print and experience is when you don't.

Contributor: Wendy Hutchinson

Even After the Heaviest Storm

Rose Kennedy

Even after the heaviest storm the birds come out singing, so why can't we delight in whatever good thing remains to us?

Contributor: George Gorman

Every Child

Theodore Roosevelt

Every child has inside him an aching void for excitement and if we don't fill it with something which is exciting and interesting and good for him, he will fill it with something which is exciting and interesting and which isn't good for him.

Contributor: Karl Rohnke

Experience through Error

Thomas Wolfe

For he had learned some of the things that every man must find out for himself, and he had found out about them as one has to find out — through error and through trial, through fantasy and illusion, through falsehood and his own damn foolishness, through being mistaken and wrong and an idiot and egotistical and aspiring and hopeful and believing and confused. As he lay there in the hospital he had gone back over his life, and, bit by bit, had extracted from it some of the hard lessons of experience. Each thing he learned was so simple and obvious, once he grasped it, that he wondered why he had not always known it. Altogether, they wove into a kind of leading thread, trailing backward through his past and out into the future. And he thought now, perhaps, he could begin to shape his life to mastery, for he felt a sense of new direction deep within him, but whither it would take him he could not say.

You Can't Go Home Again

Contributor: Mike Stratton

Exploring

Wendell Berry

Always in the big woods when you leave familiar ground and step off alone into a new place there will be, along with the feelings of curiosity and excitement, a little nagging of dread. It is the ancient fear of the Unknown, and it is your first bond with the wilderness you are going into. What you are doing is exploring. You are undertaking the first experience, not of the place, but of yourself in that place. It is an experience of our essential loneliness; for nobody can discover the world for anybody else. It is only after we have discovered it for ourselves that it becomes a common ground and a common bond, and we cease to be alone.

The One-Inch Journey

Contributor: Todd Tinkham

Fear is Nature's Warning Signal
Dr. Henry Link

Although generalizations are dangerous, I venture to say that at the bottom of most fears, both mild and severe, will be found an overactive mind and an underactive body. Hence, I have advised many people, in their quest for happiness, to use their heads less and their arms and legs more...in useful work or play. We generate fears while we sit; we overcome them in action. Fear is nature's warning signal to get busy.

Contributor: Mike Stratton

Frantic Pace
Thomas Merton

To allow one's self to be carried away by a multitude of conflicting concerns, to surrender to too many demands, to commit oneself to too many projects, to want to help everyone in everything is to succumb to violence; frenzy destroys our capacity for peace. It destroys the fruitfulness of our work, because it kills the root of inner work which makes work fruitful.

Contributor: Ted Woodward

Go For It!
Sharon Baack

Go for it!
Life lived gloriously...freely
Not sitting back and waiting
but reaching out to experience the
joys and sorrows
questions and answers
struggles and victories
and everything in between.
Go for it!
Not timidly, fearfully

inching into life's waters
but plunging in
with courage
and faith
and hope
Knowing the freedom of total commitment.
Go for it!
Not isolated...alone, but living
and learning
and growing with others
Knowing their support
and sharing it with them.
Go for it!
Rejoicing in the world around you
the people who share it...
the God who made it.
Go for it!

Contributor: Karl Rohnke

Growing Up
Aldo Leopold

When I call to mind my earliest impressions, I wonder whether the process ordinarily referred to as growing up is not actually a process of growing down; whether experience, so much touted among adults as the thing children lack, is not actually a progressive dilution of the essentials by the trivialities of living.

A Sand County Almanac

Contributor: Susan Schaefer

How Far is a Mile?
Terry and Renny Russell

How far is a mile? Well, you learn that right off. It's peculiarly different from ten tenths on the odometer. It's one thousand and seven hundred and sixty steps on the dead level and if you don't have anything better to do you can count them.

"One and a half? You're crazy, Tere, we've been walking for hours!"

It's at least ten and maybe a million times that on the hills. And no river bed ever does run straight.

"What's this, Frog Creek? Is that all the further we are? Look, tomorrow we gotta start earlier."

On the Loose

Contributor: Bob Rheault

If I Had My Life To Live Over
Nadine Stair

I'd dare to make more mistakes next time. I'd relax. I would limber up. I would be sillier than I have been this trip. I would take more chances. I would take more trips. I would climb more mountains and swim more rivers. I would eat more ice cream and less beans. I would perhaps have more troubles, but I'd have fewer imaginary ones.

You see, I'm one of those people who live sensibly and sanely hour after hour, day after day. Oh, I've had my moments, and if I had it to do over again, I'd have more of them. In fact, I'd try to have nothing else. Just moments, one after another, instead of living so many years ahead of each day. I've been one of those persons who never goes any-where without a thermometer, a hot-water bottle, a raincoat, and a parachute. If I had it to do again, I would travel lighter than I have.

If I had my life to live over, I would start barefoot earlier in the spring and stay that way later in the fall. I would go to more dances. I would ride more merry-go-rounds. I would pick more daisies.

A 78-year-old woman as quoted in the July, 1975 issue of Association for Humanistic Psychology Newsletter

Contributor: Mike Stratton

The Island
A.A. Milne

If I had a ship,
I'd sail my ship,
I'd sail my ship
Through Eastern seas;
Down to a beach were the slow waves thunder —
The green curls over and the white falls under —
Boom! Boom! Boom!
On the sun-bright sand.
Then I'd leave my ship and I'd land,
And climb the steep white sand,
And climb to the trees,
The six dark trees,
The coconut trees on the cliff's green crown —
Hands and knees
To the coconut trees,
Face to the cliff as the stones patter down,
Up, up, up, staggering, stumbling,
Round the corner where the rock is crumbling,

Round this shoulder,
Over this boulder,
Up to the top where the six trees stand...

And there would I rest, and lie,
My chin in my hands, and gaze
At the dazzle of sand below,
And the green waves curling slow,
And the grey-blue distant haze
Where the sea goes up to the sky...

And I'd say to myself as I looked so lazily down at the
    sea:
"There's nobody else in the world, and the world was
    made for me."

When We Were Very Young

Contributor: Susan Schaefer

The Joy of Climbing Well
Gaston Rebuffat

To climb smoothly between sky and earth, in a succession of precise and efficient movements, induces an inner peace and even a mood of gaiety; it is like a well-regulated ballet, with the roped climbers all in their respective places.

A difficulty encountered poses a question; the movements to resolve it give reply. This is the intimate pleasure of communicating with the mountain, not with its grandeur and beauty but, more simply, more directly, with its material self, its substance, as an artist communicates with the wood, the stone, or the iron with which he is working. There is another sort of balance, even more important than physical balance; mental balance.

The first thing is always to climb with your head. Know what you want to do and what you're capable of doing. Mountaineering is above all a question of awareness.

On Ice and Snow and Rock

Contributor: Mike Stratton

Last Words of Crowfoot
Crowfoot, a Blackfoot hunter

What is life? It is the flash of a firefly in the night. It is the breath of a buffalo in the winter time; it is the little shadow which runs across the grass and loses itself in the sunset.

Contributor: Jim Schoel

Life is Hard

Carl Sandburg

Life is hard: be steel, be a rock.
And this might stand him for the storms
And serve him for the humdrum and monotony
And guide him amid sudden betrayals
And tighten him for slack moments.

Life is soft loam; be gentle, go easy,
And this, too, might serve him.

Contributor: Bill Cuff

Look to This Day

The Sufi, 1200 B.C.

Look to this day, For it is life, the very life of life.
In its brief course lie all the varieties and realities of your
    existence;
The bliss of growth, the glory of action, The splendor of
    beauty.
For yesterday is but a dream and tomorrow is only a
    vision, But today well lived makes every yesterday a
    dream of happiness
And every tomorrow a vision of hope.
Look well, therefore, to this day, Such is the salutation of
    the dawn.

Contributor: Peter Coburn

Man In The Mirror

Author Unknown

When you get what you want in your struggle for self
And the world makes you king for a day, Then go to the
    mirror and look at yourself
And see what that guy has to say.
For it isn't your father or mother or wife
whose judgement upon you must pass
The one whose verdict counts most in life
Is that guy staring back from the glass.
He's the one to please, never mind all the rest
For he's with you clear up to the end
And you've passed your most dangerous, difficult test
If the guy in the glass is your friend.
You may be like Jack Horner and chisel a plum
And think you're a wonderful guy, But the man in the
    glass says you're a bum
If you can't look him straight in the eye.
You can fool the whole world down the path of years

And get pats on the back as you pass
But your final reward will be heartaches and tears
If you've cheated the guy in the glass.

Contributor: Jim Schoel

Meditation at Oyster River, Part 4

Theodore Roethke

Now, in this waning of light, I rock with the motion of
    morning;
In the cradle of all that is, I'm lulled into half-sleep
By the lapping of water, Cries of the sandpiper.
Water's my will, and my way, And the spirit runs,
    intermittently, In and out of the small waves, Runs
    with intrepid shorebirds—
How graceful the small before danger!
In the first of the moon, All's a scattering, A shining.

Collected Poems of Theodore Roethke

Contributor: Jim Schoel

Mother to Son

Langston Hughes

Well, son, I'll tell you:
Life for me ain't been no crystal stair.
It's had tacks in it, And splinters, And boards torn up,
    And places with no carpet on the floor —
Bare.
But all the time
I'se been a-climbin' on
And reachin' landin's,
And turnin' corners, And sometimes goin' in the dark
Where there ain't been no light.
So, boy, don't you turn back.
Don't you set down on the steps
'Cause you finds it's kinder hard.
Don't you fall now —
For I'se still goin', honey, I'se still climbin', And life for
    me ain't been no crystal stair.

Selected Poems

Contributor: Jim Schoel

Mountain Climbing

Woodrow Wilson Sayre

   The truth is that part of the essence of mountain climbing
is to push oneself to one's limits. Inevitably this involves

risk, otherwise they would not be one's limits. This is not to say that you deliberately try something you know you can't do. But you do deliberately try something which you are not sure you can do.

Four Against Everest

Contributor: Phil Salzman

Non-Metric Measurement
Dale Hayes

My life was measured in decades
And I watch as they go roaring past.
My grasp and understanding fades
In fact, I cannot recall the things
That have happened last.
I reach out to you, Through the racing stream
And the roar
Of the events
In this ascenario land
And wish that there was time for more
Than just viewing friends and happenings
As darting shades, And most particularly, that
There was some way to stop measurement
In decades.

Dr. Fossil is More Fun Than Earth People

Contributor: Bill Cuff

Our Approach: Is Falling Failing?
Mike Stratton

While you learned to walk, you fell
often…
You fell often off your first bike…
You usually fall off your first time
in a kayak…
So, when teaching climbing or just
climbing rocks…expect to fall…expect to fail…
teach how to fall…teach how to fail, how to help one who
    falls or fails (spotting and encouragement).
Falling-failing is part of climbing and of life…
The hard part is getting up and trying again
It's like knot tying…if you fail to tie the right knot, you
    untie it and try again.
You'll feel much better, safer, and more comfortable with
    the right knot.

The Bounders Bible

Contributor: Mike Stratton

Peak Achievement

Hulda Crooks, a 91-year-old mountaineer from Loma Linda, California, reached the top of Mount Fuji in Japan at dawn today after a difficult three-day climb. She is the oldest woman to conquer Mount Fuji. Crooks stepped through a special gate marking the top of the sacred 12,385-foot dormant volcano at 3:45 a.m., as a pink sun rose over the horizon and waved an American flag tied to her walking stick. "It's wonderful," she said, bundled in a down jacket at the summit in near-freezing weather. "You always feel good when you made a goal."

Contributor: Jim Schoel

Peanut Butter Sermon

HIOBS Student

To me a peanut butter and jelly sandwich after an hour or two's row is a damn luxury. A chug of water after an hour's climb up a mountain is a luxury. Jesus! You run a couple of miles and any kind of food is delicious. Yeh, and you deserve it; your body needs it and your mind as well. Use your body, push it, break it. Boy does it feel good afterwards. Use your mind, learn things, acquire new skills. Don't be afraid to make mistakes, you'll learn from them. Till your mind, work your body. They're a team, keep it in shape. Don't sit around for the rest of your life watching TV. All that does is fill your mind with crap, where better more creative thoughts could have been born. Don't waste away your life eating sundaes and driving goddamn automobiles, you'll accomplish nothing, you'll become fat and useless. You'll lose self-respect. Go out, set a goal, fight to reach it. Hurt to reach it. Wow, will you feel good when you're there. Don't run and hide from problems. Strive and suffer to overcome them. So what if you lose a little, think of what you'll have gained. When I get old and I'm lying in my death bed, I'll want to be able to look back and think: You'd never see me sitting on my ass waiting for something nice to come my way; I went out and worked and hurt and sweated for it. Jesus, was I happy when I got it.

1969

Contributor: Mike Stratton

The Quitter
Robert Service

When you're lost in the Wild, and you're scared as a
     child,
And Death looks you bang in the eye,
And you're sore as a boil, it's according to Hoyle
To cock your revolver and...die.
But the Code of a Man says: "fight all you can,"
and self-dissolution is barred.
In hunger and woe, oh, it's easy to blow...
It's the hell-served-for-breakfast that's hard.

"You're sick of the game!" Well, now, that's a shame.
You're young and you're brave and you're bright.
"You've had a raw deal!" I know — but don't squeal,
Buck up, do your damndest, and fight.
It's the plugging away that will win you the day,
So don't be a piker, old pard!
Just draw on your grit; it's so easy to quit:
It's the keeping-your-chin-up that's hard.

It's easy to cry that you're beaten — and die;
It's easy to crawfish and crawl;
But to fight and to fight when hope's out of sight —
Why, that's the best game of them all!
And though you come out of each gruelling bout,
All broken and beaten and scarred,
Just have one more try — it's dead easy to die
It's the keeping-on-living that's hard.

Collected poems of Robert Service

Contributor: Jim Schoel

Rocks and Trees
Author Unknown

Rocks and trees are part of me,
Rocks and trees are part of me.
Grass and dew are part of you,
Grass and dew are part of you.

This is the middle of the song,
It helps the others get along.
Just like the stem between the flowers and roots
You know that this is the middle of the song.

Quickly get you up and let
Your senses turn to green.
Wander till you die
And you will be there, so will I.

Contributor: Mike Stratton

Roy Eldridge
Boston Globe Editorial

"Playing was my life," Eldridge once said. "I knew the chords and I knew the melody, but I never thought about them. I'd just be in this blank space, and out the music would come." One's gratitude for what came out is exceeded only by sorrow at its source's passing.

Boston Globe, May 1, 1989

Contributor: Jim Schoel

Sailing Alone
Joshua Slocum

The acute pain of solitude experienced at first never returned. I had penetrated a mystery, and, by the way, I had sailed through a fog. I had met Neptune in his wrath, but he found that I had not treated him with contempt, and so he suffered me to go on and explore.

Sailing Alone Around the World

Contributor: Jim Schoel

Solitude is a Silent Storm
Kahlil Gibran

Solitude is a silent storm that breaks down all our dead branches. Yet it sends our living roots deeper into the living heart of the living earth. Man struggles to find life outside himself, unaware that the life he is seeking is within him.

Contributor: Jim Schoel

Solo
Mike Stratton

All I can hear is the sound of surf and the cry of gulls. Except for the islands near me, the world may have vanished as far as I know. I have heard no news from the mainland since I got to Hurricane.

I have found myself a very easy companion to live with. I have not had to deal with intense loneliness or being homesick. I have settled into my new lifestyle and now cannot imagine living any other way. I feel a great bond of friendship with the others in my watch built upon sweat and achievement and sharing. A part of me lies empty for those I love at home, but my day to go home will come. I have many miles to cover until that day.

I think the most important things to a man must be his

ability to laugh, his willingness to share and his spirit of adventure. Lose one and he becomes less of a man. Lose all three and he has no reason to live.

Everyone must live for something or someone. Some live for the wrong things, things that will satisfy their base instincts. Some live for their mate and children. Some live for the others around them. To give of oneself for another is the most beautiful action and noblest idea in existence. Life is based on sharing and brotherhood. Life will continue only so long as there is love. Without love in our lives, each of us would shrivel up and die. Some find it hard to give love. When one frees himself of his self-conscious hang-ups, then he will not be afraid to love, or cry, or laugh. What could be more natural than the feelings inside our souls? One need never be embarrassed if he is honest. One need never feel ashamed if he is true to himself. How wrong it is for people to be ashamed of what God gave them in the face of others. People try so often to rearrange themselves, to please others. Instead, say "This is the person I am," and display it proudly to those around you. People cover themselves up so much. How many of us will be able to look back at our lives and say "Yes, I have lived my life honestly and fully and I am at peace with myself as death comes." That is my goal. I may die tomorrow, so let me spend today giving as much love to those around me as I can. If people will remember me at all, I want them to say "He is a man I could turn to for help," or "He is a man on whose shoulder I could cry," or "He is a man who used understanding instead of scorn as his weapon."

These are the people I admire most. The saints on earth today are not the ones we read about or hear on TV. Today's Christs are the ghetto workers, and the storefront lawyers, and the backwoods teachers. Today's Christs work with the hungry and lost and poor — sincerely and honestly. There are 100 men who do service to get their picture in the paper for every one man who honestly gives of his life for his brothers and sisters. Only when we understand that we are all brothers and sisters will we begin to see each other in love.

Contributor: Mike Stratton

Thank You for the Nourishment
Paul Radcliffe

Most of us are so caught up in the patterns and routines of our lives that we seldom take the time to become quiet, and experience the depths within us. We are used to focusing on others, and rarely do we give ourselves the opportunity to explore the inner beauty and mystery of our own lives. When we support and nourish our own self with others, both our inner lives and our expression in the world become harmonious.

Expressing oneself is both a path to inner stillness as well as an outflow of that deeper connection with the self. In our lives, we tend to house and then to feed old tapes, illusions, and fears that serve to hide our inner beauty and block its spontaneous expression. For me, this experience (of expeditioning with close friends in small boats around Cape Breton, N.S.), and those before in Wyoming and Colorado have helped me get free from past and future illusion so that I may more clearly see who I really am. Thank you for the nourishment.

Contributor: Tim Churchard

Things Done for Gain
Author Unknown

> Things done for gain are nought
> But great things done endure.
> I ever was a fighter so I fight more
> The best and the last
> I shall hate that death bandaged my eyes and bid me creep past.
> Let me pay in a minute Life's glad arrears of pain, darkness, and cold.

Etched on a cross, McMurdo Sound in memory of three who died there waiting for Sir Ernest Shackleton's Trans-Antarctic Expedition, 1917

Contributor: Mike Stratton

To Be or Not to Be
Sam Shepherd

Shakespeare didn't mince words either. "To be, or not to be," is right to the point. You can't get much more to the point than that. That is the question. Are you going to be here or not? What is the deal? Are you going to be or not be?

Contributor: Jim Schoel

Understanding
Jiddu Krishnamurti

To understand life is to understand ourselves and that is both the beginning and end of education.

Contributor: Jim Schoel

Walk With Your Head

René Daumal

A fellow climber with more experience than I has told me, "When your feet will no longer carry you, you have to walk with your head." And it's true. Perhaps it's not in the natural order of things, but isn't it better to walk with your head than to think with your feet, as happens only too often?

Mount Analogue

Contributor: Lance Lee

When You Are Old

W.B. Yeats

When you are old and grey and full of sleep,
 And nodding by the fire, take down this book,
And slowly read, and dream of the soft look
Your eyes had once, and of their shadows deep;

How many loved your moments of glad grace,
And loved your beauty with love false or true,
But one man loved the pilgrim soul in you,
And loved the sorrows of your changing face;

And bending down beside the glowing bars,
Murmur, a little sadly, how Love fled
And paced upon the mountains overhead
And hid his face amid a crowd of stars.

Contributor: Jim Schoel

Wilderness

David Douglas

Wilderness has drawn humans closer to God throughout history. Why should we, in the twentieth century, believe this is suddenly no longer true? Long after the Exodus, in a time of recurring apostasy, Hosea spoke of God wishing to "allure" the people back into the wilderness yet again — this time to the parched hills beyond Jericho. There, wrote the prophet, God would "speak tenderly" to them.

Wilderness Sojourn

Contributor: Jay Sanderford

Wisdom
Lao Tzu

He who knows others is learned, he who knows himself is wise.

Contributor: Mike Stratton

You Cannot Stay On The Summit
René Daumal

You cannot stay on the summit forever; you have to come down again...

So why bother in the first place? Just this: what is above knows what is below, but what is below does not know what is above.

One climbs, one sees. One descends, one sees no longer but one has seen. There is an art of conducting oneself in the lower regions by the memory of what one has seen higher up.

When one can no longer see, one can at least still know.

...At the end I want to speak at length of one of the basic laws of Mount Analogue. To reach the summit, one must proceed from encampment to encampment. But before setting out for the next refuge, one must prepare those coming after to occupy the place one is leaving. Only after having prepared them can one go on up. That is why, before setting out for a new refuge we had to go back down in order to pass on our knowledge to other seekers.

Mount Analogue

Contributor: Lance Lee

Youth and the Sea
Joseph Conrad

By all that's wonderful, it is the sea, I believe, the sea itself — or is it youth alone? Who can tell? But you here — you all had something out of life: money, love — whatever one gets on the shore — and, tell me, wasn't that the best time, that time when we were young at sea; young and had nothing, on the sea that gives nothing, except hard knocks — and sometimes a chance to feel your strength — that only — what you all regret?

Youth

Contributor: Mike Stratton

LEADERSHIP

Keepers of the Trail
Grey Owl

Each succeeding generation takes up the work that is laid down by those who pass along, leaving behind them traditions and a standard of achievement that must be lived up to by those who would claim a membership in the brotherhood of the Keepers of the Trail.

Men of the Last Frontier, 1931

Contributor: Bob Henderson

Bob's comments: I use this on the last day of group canoe trips in the Canadian Shield of Ontario that was Grey Owl's favorite haunt. The passage, therefore, has a specific feel to it though it is meant to apply generally in a larger sense of bush tradition, brotherhood and trail. Most often the group forms a tight circle with our paddles touching in the center. The trip's particular spirit of goodwill is captured with jokes and rehashing of favorite events in our tight grouping. The group's distinctiveness is celebrated. Grey Owl's words provide a finale before the last portage or open stretch of water to the official end of the trip.

The message that I hope is conveyed is one of having extended oneself into this landscape's roots and traditions — to having not so much "gone on a trip" but having dwelled with the past. The group moves from rich feelings of distinctiveness to insight that is heritage rendered as vivid experience: from distinctiveness to a shared brotherhood with a sense of responsibility for the preserved integrity of the possible "fit."

A Race With Old Age

Author Unknown

A race with Old Age — long and hard with detours and distractions. But a Navajo tells her children to run towards the dawn every morning — to choose the most difficult trail because only the path that is long develops into a life worth running. A strong life. And if these young people should run well, then all good things in mind and spirit will become theirs.

Contributor: Bert Horwood

Bert's comments: Many experiential educators like to impel their students into moments of decision and choice. The tough part is that we often have strong preferences about appropriate choices for the students to make. But we violate the principles we cherish if there is not some true element of decision-making by the student. This reading is useful in advance of such a moment. It declares the value of the difficult path without removing other options.

The reading has other advantages. It can open up discussions of parental expectations. And this can be useful whether one is working with family groups, youth alone, or adults. The reading also invokes cultural diversity and offers a positive representation of a Native American view. I also like it because not many people that I work with will ever likely see the original.

Ulysses

Alfred, Lord Tennyson

It little profits that an idle king,
By this still hearth, among these barren crags,
Match'd with an aged wife, I mete and dole
Unequal laws unto a savage race,
That hoard, and sleep, and feed, and know not me.
I cannot rest from travel; I will drink
Life to the lees. All times I have enjoy'd
Greatly, have suffer'd greatly, both with those
That loved me, and alone; on shore, and when
Thro' scudding drifts the rainy Hyades
Vext the dim sea: I am become a name;
For always roaming with a hungry heart
Much have I seen and known, cities of men
And manners, climates, councils, governments,
Myself not least, but honour'd of them all,
And drunk delight of battle with my peers,
Far on the ringing plains of windy Troy.
I am a part of all that I have met;

Yet all experience is an arch wherethro'
Gleams that untravell'd world whose margin fades
For ever and for ever when I move.
How dull it is to pause, to make an end,
To rust, unburnishe'd, not to shine in use!
As tho' to breathe were life! Life piled on life
Were all too little, and of one to me
Little remains; but every hour is saved
From that eternal silence, something more,
A bringer of new things; and vile it were
For some three suns to store and hoard myself,
And this gray spirit yearning in desire
To follow knowledge, like a sinking star,
Beyond the utmost bound of human thought.

This is my son, mine own Telemachus,
To whom I leave the sceptre and the isle,
Well-loved of me, discerning to fulfill
This labor, by slow prudence to make mild
A rugged people, and thro' soft degrees
Subdue them to the useful and the good.
Most blameless is he, centered in the sphere
Of common duties, decent not to fail
In offices of tenderness, and pay
Meet adoration to my household gods,
When I am gone. He works his work, I mine.
There lies the port; the vessel puffs her sail;
There gloom the dark, broad seas. My mariners,
Souls that have toil'd, and wrought, and thought with me,
That ever with a frolic welcome took
The thunder and the sunshine, and opposed
Free hearts, free foreheads, you and I are old;
Old age hath yet his honour and his toil.
Death closes all; but something ere the end,
Some work of noble note, may yet be done,
Not unbecoming men that strove with gods.

The lights begin to twinkle from the rocks;
The long day wanes; the slow moon climbs; the deep
Moans round with many voices. Come, my friends.
'Tis not too late to seek a newer world.
Push off, and sitting well in order smite
The sounding furrows; For my purpose holds
To sail beyond the sunset, and the baths
Of all the western stars, until I die.
It may be that the gulfs will wash us down:
It may be that we shall touch the Happy Isles,
And see the great Achilles, whom we knew.
'Tho much is taken, much abides; and 'tho

We are not now that strength which in old days
Moved earth and heaven, that which we are, we are,
One equal temper of heroic hearts,
Made weak by time and fate, but strong in will
To strive, to seek, to find and not to yield.

Poems of Tennyson

Contributors: Don MacDowall and Bob Rheault

Bob's comments: To me this is the all time, great Outward Bound reading. It's #1 in my battered book of readings and was given to me by Susan St. John at Bartlett's Island on the summer solstice of 1975.

Of course, it's classic because the last lines are so close to the Outward Bound motto (To serve, to strive, and not to yield); but it has much more. It is a particularly good reading for adult groups — the older the better, when it comes to such lines as "…we are not now that strength which in old days moved earth and heaven; That which we are, we are…" and "…but something ere the end, some work of noble note, may yet be done…"

The call to adventure and to struggle is there. Don't get comfortable and fat: "How dull it is to pause, to make an end, to rust unburnished, not to shine in use!"

Like all good poetry, this is extremely rich and hard to absorb all at once. I like to use it at, or close to, the end of the course and give copies of it to departing students or send it in a follow-up letter urging all not to "store and hoard" themselves but to "push off" and "seek a newer world" and "to strive, to seek, to find and not to yield."

Aikido in Action
Terry Dobson

The train clanked and rattled through the suburbs of Tokyo on a drowsy spring afternoon. Our car was comparatively empty…a few housewives with their kids, some old folks going shopping. I gazed absently at the drab houses and dusty hedgerows.

At one station the doors opened, and suddenly the afternoon quiet was shattered by a man bellowing violent, incomprehensible curses. The man staggered into our car. He wore laborer's clothing, and he was big, drunk and dirty. Screaming, he swung at a woman holding a baby. The blow sent her spinning into the laps of an elderly couple. It was a miracle that the baby was unharmed.

Terrified, the couple jumped up and scrambled toward the other end of the car. The laborer aimed a kick at the retreating back of the old woman but missed as she scuttled to safety. This so enraged the drunk that he grabbed the metal pole in the center of the car and tried to wrench it out of its stanchion. I could see that one of his hands was cut and bleeding. The train lurched ahead, the passengers frozen with fear. I stood up.

I was young then, some twenty years ago, and in pretty good shape. I'd been putting in a solid eight hours of aikido training nearly every day for the past three years. I liked to throw and grapple. I thought I was tough. Trouble was, my martial skills were untested in actual combat. As students of aikido, we were not allowed to fight.

"Aikido," my father said again and again, "is the art of reconciliation. Whoever has the mind to fight has broken his connection with the universe. If you try to dominate people, you are already defeated. We study how to resolve conflict, not how to start it."

I listened to his words, I tried hard. I even went so far as to cross the street to avoid the Chimpira, the pinball punks who lounged around the train stations. My forbearance exalted me. I felt both tough and holy. In my heart, however, I wanted an absolutely legitimate opportunity whereby I might save the innocent by destroying the guilty.

This is it! I said to myself as I got to my feet. People are in danger. If I don't do something fast, somebody will probably get hurt. Seeing me stand up, the drunk recognized a chance to focus his rage. "Aha!" he roared. "A foreigner! You need a lesson in Japanese manners!" I held on lightly to the commuter strap overhead and gave him a slow look of disgust and dismissal. I planned to take this turkey apart, but he had to make the first move. "All right!" he hollered. You're gonna get a lesson." He gathered himself for a rush at me.

A split second before he could move, someone shouted "Hey!" It was earsplitting. I remember the strangely joyous, lilting quality of it, as though you and a friend had been searching diligently for something and he had suddenly stumbled upon it. "Hey!"

I wheeled to my left; the drunk spun to his right. We both stared down at a little old Japanese. He must have been well into his seventies, this tiny gentleman, sitting there immaculate in his kimono. He took no notice of me but beamed delightedly at the laborer, as though he had a most important most welcome secret to share.

"C'mere," the old man said in an easy vernacular, beckoning to the drunk. "C'mere and talk with me." He waved his hand lightly.

The big man followed as if on a string. He planted his feet belligerently in front of the old gentleman and roared above the clacking wheels, "Why the hell should I talk to you?" The drunk now had his back to me. If his elbow moved so much as a millimeter, I'd drop him in his socks.

The old man continued to beam at the laborer. "What'cha been drinkin?" he asked, his eyes sparkling with interest. "I been drinkin sake," the laborer bellowed back, "and it's none of your business!" Flecks of spittle spattered the old man. "Oh, that's wonderful," the old man said, "absolutely wonderful! You see, I love sake too. Every night, me and my wife (she's seventy-six, you know), we warm up a little bottle of sake and take it out into the garden, and we sit on an old wooden bench. We watch the sun go down, and we look to see how our persimmon tree is doing. My great-grandfather planted that tree, and we worry about whether it will recover from those ice storms we had last winter. Our tree has done better than I expected, though, especially when you consider the poor quality of the soil. It is gratifying to watch when we take our sake and go out to enjoy the evening...even when it rains." He looked up at the laborer, eyes twinkling.

As he struggled to follow the old man's conversation, the drunk's face began to soften. His fists slowly unclenched. "Yeah," he said, "I love persimmons too..." His voice trailed off. "Yes," said the old man, smiling, "And I'm sure you have a wonderful wife."

"No," replied the laborer. "My wife died." Very gently, swaying with the motion of the train, the big man began to sob. "I don't got no wife. I don't got no home. I don't got no job. I'm so ashamed of myself." Tears rolled down his cheeks; a spasm of despair rippled through his body.

Now it was my turn. Standing there in my well-scrubbed youthful innocence, my make-this-world-safe-for-democracy righteousness, I suddenly felt dirtier than he was.

Then the train arrived at my stop. As the doors opened, I heard the old man cluck sympathetically. "My, my," he said. "That is a difficult predicament, indeed. Sit down here and tell me about it." I turned my head for one last look. The laborer was sprawled on the seat, his head in the old man's lap. The old man was softly stroking the filthy, matted hair.

As the train pulled away I sat down on a bench. What I had wanted to do with muscle had been accomplished with kind words. I had just seen Aikido tried in combat, and the essence of it was love. I would have to practice the art with an entirely different spirit. It would be a long time before I could speak about the resolution of conflict.

Unitarian Universalist "World"

Contributor: Jim Schoel

Being a Midwife

Lao Tzu

The wise leader does not intervene unnecessarily. The leader's presence is felt, but often the group runs itself.

Lesser leaders do a lot, say a lot, have followers, and form cults. Even worse ones use fear to energize the group and force to overcome resistance.

Only the most dreadful leaders have bad reputations. Remember that you are facilitating another person's process. It is not your process. Do not intrude. Do not control. Do not force your own needs and insights into the foreground.

If you do not trust a person's process, that person will not trust you. Imagine that you are a midwife; you are assisting at someone else's birth. Do good without show or fuss. Facilitate what is happening rather than what you think ought to be happening. If you must take the lead, lead so that the mother is helped, yet still free and in charge.

When the baby is born, the mother will rightly say: "We did it ourselves!"

The Tao of Leadership: Lao Tzu's Tao Te Ching Adapted for
a New Age

Contributor: Laurel Hood

Children

Kahlil Gibran

Your children are not your children.
They are the sons and daughters of Life's longing for itself.
They come through you but not from you,
And though they are with you yet they belong not to you.

You may give them your love but not your thoughts,
For they have their own thoughts.
You may house their bodies but not their souls,
For their souls dwell in the house of tomorrow, which you cannot visit, not even in your dreams.
You may strive to be like them, but seek not to make them like you,
For life goes not backward nor tarries with yesterday.

You are the bows from which your children as living arrows are sent forth
The archer sees the mark upon the path of the infinite, and He bends you with His might that His arrows may go swift and far.

Let your bending in the archer's hand be for gladness;

For even as he loves the arrow that flies, so he loves also the bow that is stable.

The Prophet

Contributor: Jim Schoel

Children Learn What They Live
Author Unknown

If a child lives with criticism, he learns to condemn.

If a child lives with hostility, he learns to fight.

If a child lives with fear, he learns to be apprehensive.

If a child lives with pity, he learns to feel sorry for himself.

If a child lives with ridicule, he learns to be shy.

If a child lives with jealousy, he learns what envy is.

If a child lives with shame, he learns to feel guilty.

If a child lives with encouragement, he learns to be confident.

If a child lives with tolerance, he learns to be patient.

If a child lives with praise, he learns to be appreciative.

If a child lives with acceptance, he learns to love.

If a child lives with approval, he learns to like himself.

If a child lives with recognition, he learns that it is good to have a goal.

If a child lives with sharing, he learns about generosity.

If a child lives security, he learns to have faith in himself and in those about him.

If you live with serenity, your child will live with peace of mind!

Contributor: Wendy Hutchinson

Decisive Element in Classroom
Haim Ginott

I've come to a frightening conclusion that I am the decisive element in the classroom. It is my personal approach that creates the climate. It is my daily mood that makes the weather. As a teacher, I possess a tremendous power to make a child's life miserable or joyous. I can be a tool of torture or an instrument of inspiration. I can humiliate or humor, hurt or heal. In all situations it is my response that decides whether a crisis will be escalated or de-escalated and a child humanized or de-humanized.

Teacher & Child, preface

Contributor: Howie Futterman

Do Not Handicap
Robert Heinlein

Do not handicap your children by making their lives easy.

The Notebooks of Lazarus Long

Contributor: Karl Rohnke

Duty of Living
Dag Hammarskjöld

Your position never gives you the right to command. It only imposes on you the duty of living your life that others can receive your orders without being humiliated.

Markings

Contributor: Jim Schoel

The Educational Engineer
Leon Lessinger

A good engineer begins by challenging assumptions. He refuses to believe that something is impossible merely because it has never been done or because people say there is no way to do it or because it would upset established ways. The good engineer, in the field of education as elsewhere, starts with a goal to be achieved, not with the dead weight of precedent or unexamined beliefs. Like the runner who finally broke the four minute mile, he knows that the limits of possibility are stretched not solely by pushing from within, but by setting an outside goal and doing what is necessary to reach it. In track competition, once the four minute mark was set, other runners soon matched the feat, mainly because they had learned that it was possible. Their assumptions had changed.

Every Kid a Winner

Contributor: Pat Sheckler

Good Guide
Gaston Rebuffat

This profession might become wearisome through the repetition of the same climbs time after time, but the guide is more than a mere machine for climbing rocks and ice slopes, for knowing the weather and the way.

He knows that such-and-such a climb is particularly interesting, that at this turn the view is quite suddenly very beautiful, and that this ice ridge is delicate as lace. He says nothing of all this but his reward is in his companion's smile of discovery.

Starlight and Storm

Contributor: Rich Obenschain

A Handful of Men
Henry Miller

It nevertheless remains an illuminating fact that it is only the presence of a handful of men, in every age, that keeps society from degenerating utterly.

The Hour of Man

Contributor: Todd Tinkham

The High School Teacher
Naomi J. White

I have taught in high school for ten years. During that time I have given assignments, among others, to a murderer, an evangelist, a pugilist, a thief, and an imbecile.

The murderer was a quiet little boy who sat on the front seat and regarded me with pale blue eyes; the evangelist, easily the most popular boy in school, had the lead in the Junior play; the pugilist lounged by the window and let loose at intervals a raucous laugh that startled even the geraniums; the thief was a gay-hearted Lothario with a song on his lips; and the imbecile, a soft-eyed little animal seeking the shadows.

The murderer awaits death in the state penitentiary, the evangelist has lain a year now in the village churchyard; the pugilist lost an eye in a brawl in Hong Kong; the thief, standing on tiptoe, can see the windows of my room from the county jail; and the once gentle-eyed little moron beats his head against a padded wall in the state asylum.

All of these pupils once sat in my room, sat and looked at me gravely across worn-down desks. I must have been a great help to those pupils — I taught them the rhyming scheme of the Elizabethan sonnet and how to diagram a complex sentence.

Contributor: Jim Schoel

I Went on a Search
Author Unknown

I went on a search to become a leader. I searched high and low. I spoke with authority, people listened. But at last there was one who was wiser than I and they followed him. I sought to inspire confidence but the crowd responded, "Why should we trust you?" I postured and I assumed the look of leadership with a countenance that glowed with confidence and pride. But the crowd passed by and never noticed my air of elegance. I ran ahead of the others pointing new ways to new heights. I demonstrated that I knew the route to greatness. And then I looked back and I was alone. "What shall I do?" I queried. "I've tried hard and used all that I know." And then I listened to the voices around me. And I heard what the group was trying to accomplish. I rolled up my sleeves and joined in the work. As we worked I asked, "Are we all together in what we want to do, and how we'll get the job done?" And we thought together and we struggled towards our goal. I found myself encouraging the faint hearted. I sought the ideas of those too shy to speak out. I taught those who knew little at all. I praised those who worked hard. When our task was completed, one of the group members turned to me and said, "This would not have been done but for your leadership." At first I said, "I did not lead, I just worked with the rest." And then I understood — leadership isn't a goal. I lead best when I forget about myself as a leader and focus on my group, their needs and their goals. To lead is to serve, to give, to achieve together.

Contributor: Ted Woodward

If —
Rudyard Kipling

> If you can keep your head when all about you
> Are losing theirs and blaming it on you,
> If you can trust yourself when all men doubt you,
> But make allowance for their doubting too;
> If you can wait and not be tired by waiting,
> Or being lied about, don't deal in lies,
> Or being hated don't give way to hating,
> And yet don't look too good, nor talk too wise:
>
> If you can dream — and not make dreams your master,
> If you can think— and not make thoughts your aim;
> If you can meet with Triumph and Disaster
> And treat those two imposters just the same;
> If you can bear to hear the truth you've spoken
> Twisted by knaves to make a trap for fools,
> Or watch the things you gave your life to broken,
> And stoop and build em up with worn-out tools:
>
> If you can make one heap of all your winnings
> And risk it on one turn of pitch and toss,
> And lose, and start again at your beginnings

And never breathe a word about your loss;
If you can force your heart and nerve and sinew
To serve your turn long after they are gone,
And so hold on when there is nothing in you
Except the Will which says to them, "Hold on."

If you can talk with crowds and keep your virtue,
Or walk with kings — nor lose common touch,
If neither foes nor loving friends can hurt you,
If all men count with you, but none too much;
If you can fill the unforgiving minute
With sixty seconds' worth of distance run,
Yours is the Earth and everything that's in it,
And — which is more, you'll be a Man, my son.

Contributor: Mike Stratton

It's Not the Critic Who Counts
Theodore Roosevelt

It's not the critic who counts. Not the man who points out where the strong man stumbled or where the doer of great deeds could have done them better. The credit belongs to the man who is actually in the arena. Whose face is marred by dust and sweat and blood. Who strives valiantly, who errs and comes up short again and again. And who, while daring greatly, spends himself in a worthy cause so that his place may never be among those cold and timid souls who know neither victory nor defeat.

While Daring Greatly

Contributor: Mike Stratton

Learning to Read
Garrison Keillor

It took me a long time to learn to read. I was wrong about so many words. *Cat, can't. Tough, through, thought. Shinola.* It was like reading a cloud of mosquitoes. Donna in the seat behind whispered right answers to me, and I learned to be a good guesser, but I didn't read well until Mrs. Meiers took me in hand.

One winter day she took me aside after recess and said she'd like me to stay after school and read to her. "You have such a nice voice," she said, "and I don't get to hear you read in school as much as I'd like."

No one had told me before that I had a nice voice. She told me many times over the next few months what a *wonderful* voice I had, as I sat in a chair by her desk reading to her as she marked worksheets. "The little duck was so happy. He ran to the barn and shouted, 'Come! Look! The ice is gone from the pond!' Finally it was spring."

"Oh, you read that so well. Read it again," she said. When Bill the janitor came in to mop, she said, "Listen to this. Doesn't this boy have a good voice?" He sat down and I read to them both. "The little duck climbed to the top of the big rock and looked down at the clear blue water. 'Now I am going to fly,' he said to himself. He waggled his wings and counted to three. 'One, two, three.' And he jumped and..." I read in my clear blue voice. "I think you're right," Bill said. "I think he has a very good voice. I wouldn't mind sitting here all day and listening to him."

Lake Wobegon Days

Contributor: Susan Schaefer

Look For a Reaction
Jemique Straker

Students are young and they will live their lives. They can't live off of someone else's experiences. So as you try to prepare them don't look for an answer; try looking for a reaction.

Contributor: Jemique Straker

Nature as Teacher
Saint Bernard of Clairvaux

What I know of the Divine sciences and Holy Scriptures, I learnt in woods and fields. I have had no other masters than the beeches and the oaks. Listen to a man of experience; thou wilt learn more in the woods than thou canst acquire from the mouth of a magister.

Contributor: Ted Woodward

Room for Growth
Author Unknown

A leader creates space that:
- empowers others.
- inspires others.
- makes conscious that which is unconscious in others.

Contributor: Peter Coburn

Souls of Black Folk
Dr. W.E.B. Dubois

Men of America, the problem is plain before you. Here is a race transplanted through the criminal foolishness of your fathers. Whether you like it or not the millions are here, and here they will remain. If you do not lift them up, they will pull you down. Education and work are the levers to uplift a people. Work will not do it unless inspired by the right ideals and guided by intelligence. Education must not simply teach work — it must teach life. The talented tenth of the Negro race must be made leaders of thought and missionaries of culture among their people. No others can do this work and the Negro colleges must train men for it. The Negro race, like all other races, is going to be saved by its exceptional men.

Souls of Black Folk

Contributor: Arthur Conquest III

Success
Ralph Waldo Emerson

To laugh often and much; to win the respect of intelligent people and affection of children; to earn the appreciation of honest critics and endure the betrayal of false friends; to appreciate beauty, to find the best in others; to leave the world a bit better, whether by a healthy child, a garden patch or a redeemed social condition; to know even one life has breathed easier because you have lived. This is to have succeeded.

Contributor: Beau Bassett

Three Ways
Kurt Hahn

There are three ways of trying to win the young. There is persuasion, there is compulsion and there is attraction. You can preach at them; that is a hook without a worm. You can say "You must volunteer." That is of the devil. And you can tell them, "You are needed." That hardly ever fails.

Contributor: Mike Stratton

VALUES

Becoming Real
Margery Williams

The Skin Horse had lived longer in the nursery than any of the others. He was so old that his brown coat was bald in patches and showed the seams underneath, and most of the hairs in his tail had been pulled out to string bead necklaces. He was wise, for he had seen a long succession of mechanical toys arrive to boast and swagger, and by-and-by break their springs and pass away, and he knew that they were only toys, and would never turn into anything else. For nursery magic is very strange and wonderful, and only those playthings that are old and wise and experienced like the Skin Horse understand all about it.

"What is REAL?" asked the Rabbit one day, when they were lying side by side near the nursery fender, before Nana came to tidy the room. "Does it mean having things that turn inside you and a stick-out handle?"

"Real isn't how you are made," said the Skin Horse. "It's a thing that happens to you. When a child loves you for a long, long time, not just to play with, but REALLY loves you, then you become Real."

"Does it hurt?" asked the Rabbit.

"Sometimes," said the Skin Horse, for he was always truthful. "When you are Real you don't mind being hurt."

"Does it happen all at once, like being wound up," he asked, "or bit by bit?"

"It doesn't happen all at once," said the Skin Horse. "You become. It takes a long time. That's why it doesn't often happen to people who break easily, or have sharp edges, or who have to be carefully kept. Generally, by the time you are Real, most of your hair has been loved off, and your eyes drop out and you get loose in the joints and very shabby. But these things don't matter at all, because once you are Real you can't be ugly, except to people who don't understand."

The Velveteen Rabbit, or How Toys Become Real

Contributor: Rafe Parker

Rafe's comments: This is such a powerful reading. It was read to me by a student on a teacher's course on Bartlett's Island in the summer of 1970. It never ceases to bring tears to my eyes and has always been a favorite on adult programs. I read it once on a managers course while running the Rio Grande in Big Bend. The Velveteen Rabbit became such an important symbol on the course that at the end of it one of the students sent everyone else a toy velveteen rabbit and a copy of the book.

One Must Look With the Heart
Antoine de Saint-Exupéry

I raised the bucket to his lips. He drank, his eyes closed. It was as sweet as some special festival treat. This water was indeed a different thing from ordinary nourishment. Its sweetness was born of a walk under the stars, the song of the pulley, the effort of my arms. It was good for the heart, like a present. When I was a little boy, the lights of the Christmas tree, the music of the Midnight Mass, the tenderness of smiling faces, used to make up the radiance of the gifts I received.

"The men where you live," said the little prince, "raise five thousand roses in the same garden — and they do not find in it what they are looking for."

"They do not find it." I replied.

"And yet what they are looking for could be found in one single rose, or in a little water."

"Yes, that is true," I said.

And the little prince added:

"But the eyes are blind. One must look with the heart…"

The Little Prince

Contributor: Susan Schaefer

Susan's comments: There is no complex pantheistic philosophy here, but simply an observation on the wonder of Creation, and the fact that we appreciate most the things we have lost.

The Sea
Gordon Bok

The sea takes trouble from you; takes worry and fear and illusion and anger and joy and joking and plans and ambition and love from you. Takes them, scatters them, gathers them, gives them back to you not so big or important as before.

You're not anyone, really; you never were. Oh you

thought you were, when your head was too small for your illusions. But illusions aren't important now: you don't have to be anything, even yourself, because yourself was only something you had to make up, and then you thought you had to carry it around with you. What a relief to lay it down and walk away and forget it. Just to be a part of what's around you is enough.

Another Land Made of Water

Contributor: Peter Coburn

Peter's comments: So much of Outward Bound is DOING: running, reaching, beating, jumping, climbing, dipping, hiking, rowing, cooking, writing, watching…The list seems endless. The purpose of all this activity, it seems to me, is to shake us out of our patterns, our habits, our long-accustomed roles. Now this is all very worthwhile and for some in Outward Bound, that's all there is. But beyond all this DOING is the possibility of BEING and that is what this reading is about for me.

The reading puts me in touch with what Willi Unsoeld called the fourth concentric circle of Outward Bound, those difficult-to-describe matters of the spirit, the heart, and the universe. I use this reading when I think it time to shift some of the focus of the course from doing to being. This tends to be towards the end of the course: before solo and before or during final expedition.

"Dunk 'em and dry 'em," said the master. Perhaps this reading can provide a little contrast to give better perspective to our so-active programs.

We Must Learn To Reawaken
Henry David Thoreau

We must learn to reawaken and keep ourselves awake, not by mechanical aids, but by an infinite expectation of the dawn, which does not forsake us in our soundest sleep. I know of no more encouraging fact than the unquestionable ability of man to elevate his life by a conscious endeavor. It is something to be able to paint a particular picture, or to carve a statue, and so to make a few objects beautiful; but it is far more glorious to carve and paint the very atmosphere and medium through which we look, which morally we can do. To affect the quality of the day, that is the highest of arts. Every man is tasked to make his life, even in its details, worthy of the contemplation of his most elevated and critical hour.

Walden

Contributor: Rafe Parker

Rafe's comments: This old favorite never ceases to remind me of the very simple goal that each of us has, "To affect the very quality of the day." And yet, so often, it is the highest challenge that we have to face as another day begins. I always remember the African O.B. students in central Africa getting a great kick out of it. They were always very good at "painting the very atmosphere through which we look," and they would enjoy taking me to task when I was taking things too seriously.

Letter to Danny
Bill Cuff

To Little Danny:

We approached you with sadness and you gave us back joy.

We wished that you could move like us, but you are so much closer to our destination.

We hoped that you could speak like us but you have spoken so much more profoundly.

We wanted you to be more like us and now we want to be more like you.

We call you Little Danny, but we don't have your vision.

Each time we say Little Danny, let it be a reminder of how much harder we must work with our shallow eyes to see the truth.

Love, Bill

Contributor: Tim Churchard

Tim's comments: Little Danny, who is not so little now, has been totally disabled since birth. His father is a participant in a men's group along with Bill Cuff, the author of this letter and a Project Adventure Trainer. Danny is the focal point of "Danny's Team," a group of people around Rochester and Durham, New Hampshire who are dedicated to providing research and services to persons who are "challenged" in one form or another.

What Are the Qualities of Life?
Mike Gass

What are the qualities of Life?

It can be shared forever, but cannot be kept forever.

When given to another, it brings great joy to all; when taken, the anguish for many is great.

It is sometimes maintained by less than the thread of a string and sometimes lost despite the hope of millions.

Its frailty and end are obvious, but its strength and limits endless.

So share your life with me while we are together so we can create that which will bring joy to others.

When this life is over, let the meaning of our lives be found not on a list of accomplishments, but in the hearts and souls of the people with whom we shared our fragile existence.

Let our lives not be measured so much by what we did for others, but by what we helped people do for themselves.

June 1982

Contributor: Mike Gass

Mike's comments: In June of 1982, I was on a course with Rocky Kimball and other mental health professionals doing training in the use of adventure experiences for therapy. We were on the 5th day of the course and were hiking at 12,000 ft. in the Pecos Wilderness area (near Mt. Baldy, I believe). We had stopped in a valley near a pristine mountain lake when one of our members became extremely ill with internal injuries. We found out later that an ulcer had erupted. That night, her vital signs fell dangerously low, incapacitating her from moving. Pulling her out of our mountainous spot was difficult, because of the altitude and distance to the nearest phone.

Early that morning, Rocky and I started running out of the wilderness to get help while the rest of the group littered her down to 9,000 feet so that a rescue helicopter could land and pick her up to transport her to a hospital. We ended up running about 21 miles, finally getting a ride to a local town and phone. We called for rescue support, and Rocky flew back in the helicopter to point out where the evacuation spot was while I rode in a police car to Santa Fe hospital where our group member was being flown. I arrived at the same time she did (it's truly amazing how fast helicopters fly!) and sat down in the intensive care unit with her. At the same time, another person was brought in with a similar condition and placed in the same room. Both had IV's placed in their arms and both of their vital signs were at dangerous levels. As the day wore on, our party member improved while the other patient worsened and died from internal injuries.

After the experience, I hiked back in to join up with the group. During a 24-hour solo experience the next day, I contemplated the experience and life's strengths, weaknesses and meaning. Literally five minutes before the solo ended, I wrote this poem and have found it to be quite inspirational for a variety of people in looking at the "true" meaning and value of life.

A.E.F. Doughboy Prayer
Anonymous

The supreme prayer of my heart is not to be rich, famous, powerful, or "too good," but to be radiant. I desire to radiate health, calm courage, cheerfulness and good will. I wish to live without hate, whim, jealousy, envy or fear. I wish to be simple, honest, frank, natural, clean in mind and clean in body, unaffected, ready to say "I do not know," if so it be, to meet all men on an absolute equality, to face any obstacle and meet every difficulty unabashed and unafraid. I wish others to live their lives, too, up to their fullest and best. To that end I pray that I may never meddle, interfere, dictate, give advice that is not wanted, or assist when my services are not needed. If I can help people I will do it, by giving them a chance to help themselves; and if I can uplift or inspire, let it be by example, inference, suggestion, rather than by injunction and dictation. That is to say, I desire to be radiant, to radiate life.

> *Prayer found near Tours, France, in 1918 by A.E.F.*
> *Doughboy of World War I*

Contributor: Jim Schoel

A Kiss of the Sun
Author Unknown

> A kiss of the sun for pardon
> The song of the sun for mirth
> One is nearer God's heart in a garden
> Than anywhere else on earth.

Contributor: Nicole Richon-Schoel

All I Ever Really Needed to Know I Learned in Kindergarten
Robert Fulghum

Most of what I really need to know about how to live, and what to do, and how to be, I learned in kindergarten. Wisdom was not at the top of the graduate school mountain, but there in the sandbox at nursery school.

These are the things I learned: Share everything. Play fair. Don't hit people. Put things back where you found them. Clean up your own mess. Don't take things that aren't yours. Say you're sorry when you hurt somebody. Wash your hands before you eat. Flush. Warm cookies and cold milk are good for you. Live a balanced life. Learn some and think some and draw and paint and sing and dance and play and work every day some.

Take a nap every afternoon. When you go out into the world, watch for traffic, hold hands and stick together. Be aware of wonder. Remember the little seed in the plastic cup. The roots go down and the plant goes up and nobody really knows how or why, but we are all like that.

Goldfish and hamsters and white mice and even the little seed in the plastic cup — they all die. So do we.

And then remember the book about Dick and Jane and the first word you learned, the biggest word of all: LOOK. Everything you need to know is in there somewhere. The Golden Rule and love and basic sanitation. Ecology and politics and sane living.

Think of what a better world it would be if we all — the whole world — had cookies and milk about 3 o'clock every afternoon and then lay down with blankets for a nap. Or if we had a basic policy in our nation to always put things back where we found them and cleaned up our own messes. And it is still true, no matter how old you are, when you go out into the world, it is best to hold hands and stick together.

Contributors: Brian Pritchard and Jim Moll

Alone in the Wilderness
Alan Watts

To spend a lengthy period alone in the forests or mountains, a period of coming to terms with the solitude and nonhumanity of nature is to discover who, or what, one really is — a discovery hardly possible while the community is telling you what you are, or ought to be.

Nature, Man and Woman

Contributor: Mike Stratton

Armies of the Mindless
Kenneth Rexroth

Against the armies of the mindless I will take what few allies I can find, whatever their faults.

Contributor: Todd Tinkham

Beauty
Anonymous Shaker source

All beauty that has not a foundation in use, soon grows distasteful, and needs continual replacement with something new.

Contributor: Jim Schoel

Be Tough, Yet Gentle

Author Unknown

Be tough, yet gentle, Humble, but bold, Swayed always by Beauty and Truth.

Used by Bob Pieh on Minnesota Outward Bound letterhead

Contributor: Mike Stratton

Cape Horn Christmas

E.R. Warren

Storm canvas set to face the Greybeards' might:
The main with goosewing lower top, a fore
Storm stays'l, and a trysail reefed down tight,
And extra gaskets round the jibs. No more
Can now be done by sailormen but wait
And hold hard on a course of south-southwest
Where mammoth seas, strong ships, ice, wind join fate.
Equilibrium. Manhood's hardest test.
The watch is called: hot coffee and hard bread
Our Christmas breakfast, honey-laced to fete
The Holy Babe; while we, like living dead
Still oilskinned, cling to cabin's lee — not yet
Awake, yet smiling as we face this morn,
Our childhood hopes again renewed. The Horn!

Contributor: Mike Stratton

Central Theme of Existentialism

Gordon W. Allport

To live is to suffer — to survive is to find meaning in the suffering. If there is a purpose in life at all, there must be a purpose in suffering and in dying. But no man can tell another what this purpose is. Each must find out for himself, and must accept the responsibility that his answer pre-scribes. If he succeeds he will continue to grow in spite of all indignities.

Preface to Man's Search for Meaning

Contributor: Jim Schoel

Dandelion Wine

Ray Bradbury

"Tom," said Douglas, "Just promise me one thing, okay?"
"It's a promise. What?"
"You may be my brother and maybe I hate you some-times, but stick around, all right?"

"You mean you'll let me follow you and the older guys when you go on hikes?"

"Well...sure...even that. What I mean is, don't go away, huh? Don't let any cars run over you or fall off a cliff."

"I should say not! Whatta you think I am, anyway?"

"'Cause if worst comes to worst, and both of us are real old — say forty-five some day — we can own a gold mine out west and sit there smoking corn silks and growing beards."

"Growing beards! Boy!"

"Like I say, you stick around and don't let nothing happen."

"You can depend on me," said Tom.

"It's not you I worry about," said Douglas. "It's the way God runs the world."

Tom thought about this for a moment.

"He's all right, Doug," said Tom. "He tries."

Dandelion Wine

Contributor: Dr. Richard Maizell

Different Drummer
Henry David Thoreau

Why should we be in such haste to succeed and in such desperate enterprises? If a man does not keep pace with his companions, perhaps it is because he hears a different drummer. Let him step to the music which he hears, however measured or far away.

Walden

Contributor: Scott Garman

Disposable Society
Eugene McDaniels

Disposable society has thrown away the best in me.
They've thrown away sincerity.
The keystone of integrity.
Disposable to throw away.
Buy something new another day.
There's nothing new that's made to stay.
Planned obsolescence will make you pay.

Paper plates, cardboard skates, plastic silverware,
Automobiles with disposable wheels,
Wigs instead of hair.
That's how it is.

Disposable way you love,
Not exactly what you're thinking of.
Dispose of me when you are through
For fear that I'll dispose of you.
Disposable your closest friend.
You swore you'd love right to the end.
Your rigid mind won't let you bend.
You're further gone than you pretend.

Contributor: Jim Schoel

Do Not Pray
John F. Kennedy

Do not pray for easy times; pray to be stronger. Do not pray for tasks equal to your powers; pray for powers equal to your tasks.

Contributor: Jim Schoel

Don't Look Back
Satchel Paige

Avoid fried meats which angry up the blood.

If your stomach disputes you, lie down and pacify it with cool thoughts.

Keep the juices flowing by jangling around gently as you move.

Go very light on the vices such as carrying on in society; the social rumble ain't restful.

Avoid running at all times.

Don't look back; something might be gaining on you.

Contributor: Jim Schoel

Education
Robert Frost

Education is the ability to listen to almost anything without losing your temper or your self-control.

Contributor: Beau Bassett

Every Man's Destiny
Henry Miller

Every man has his own destiny: the only imperative is to follow it, to accept it, no matter where it lead him.

The Wisdom of the Heart

Contributor: Todd Tinkham

Get Out of That Rut

Author Unknown

Oscar Wilde said,
"Consistency is
the last refuge of
the unimaginative."
So stop getting up
at 6:05.
Get up at 5:06.
Walk a mile at dawn.
Find a new way
to drive to work.
Switch chores with
your spouse
next Saturday.
Buy a wok.
Study wildflowers.
Stay up alone all night.
Read to the blind.
Start counting
brown-eyed blondes
or blonds.
Subscribe to an
out-of-town paper.
Canoe at midnight.
Don't write to your
congressman, take a whole scout
troop to see him.
Learn to speak
Italian.
Teach some kid
the thing you do best.
Listen to two hours of
uninterrupted Mozart.
Take up aerobic dancing.
Leap out of that rut.
Savor life.
Remember, we only
pass this way once.

Contributor: Howie Futterman

Give Me a Good Digestion

Author Unknown

Give me a good digestion, Lord,
And also something to digest.
Give me a healthy body, Lord,
With a sense to keep it at its best.

Give me a healthy mind, Good Lord,
To keep the good and pure in sight,
Which seeing sin is not appalled
But finds a way to put it right.
Give me a mind that is not bored
And does not whimper, whine or sigh.
Don't let me worry overmuch
About that fussy thing called "I."
Give me a sense of humor, Lord,
Give me the grace to see a joke,
To get some happiness from life
And pass it on to other folk.

Found in Winchester Cathedral, c. 15th century

Contributor: Jim Schoel

Hamlet
William Shakespeare

This above all: to thine own self be true,
And it must follow, like the night the day
thou canst not be false to any man.

Hamlet

Contributor: Mike Stratton

I Am Glad I Did It
Mark Twain

I am glad I did it, partly because it was well worth it, and
chiefly because I shall never have to do it again.

Huckleberry Finn

Contributor: Mike Stratton

I Am Not Bound to Win
Abraham Lincoln

I am not bound to win, but I am bound to be true.
I am not bound to succeed, but I am bound to live up to
what light I have.

Contributor: Jim Schoel

I Asked God
An Unknown Confederate Soldier

I asked God for strength, that I might achieve;
I was made weak, that I might learn humbly to obey.
I asked for health, that I might do greater things;

I was given infirmity, that I might do better things.
I asked for riches, that I might be happy;
I was given poverty, that I might be wise.
I asked for power, that I might have praise of men;
I was given weakness, that I might feel the need of God.
I asked for all things, that I might enjoy life, I was given
   life, that I might enjoy all things.
I got nothing that I asked for, but everything I had hoped
   for;
Almost despite myself, my unspoken prayers were
   answered.
I am among all men, most richly blessed.

Contributor: Susan Schaefer

I Enjoy More Drinking Water
Henry David Thoreau

I enjoy more drinking water at a clear spring than out of a
goblet at a gentleman's table. I like best the bread which I
have baked, the garment which I have made, the shelter
which I have constructed, the fuel which I have gathered.

Contributor: Mike Stratton

I Went to the Woods
Henry David Thoreau

I went to the woods because I wished to live deliberately,
to front only the essential facts of life, and see if I could not
learn what it had to teach, and not, when I came to die,
discover that I had not lived. I did not wish to live what was
not life, living is so dear; nor did I wish to practice resignation, unless it was quite necessary. I wanted to live deep and
search out all the marrow of life, to live so sturdily and
Spartan-like as to put to rout all that was not life, to cut a
broad swath and shave close, to drive life into a corner, and
reduce it to its lowest terms, and, if it proved to be mean,
why then to get the whole and genuine meanness of it, and
publish its meanness to the world; or if it were sublime, to
know it by experience, and be able to give a true account of
it in my next excursion.

Walden, or Life in the Woods

Contributor: Mike Stratton

I Left the Woods
Henry David Thoreau

I left the woods for as good a reason as I went there. Perhaps it seemed to me that I had several more lives to live, and could not spare any more time for that one. It is remarkable how easily and insensibly we fall into a particular route, and make a beaten track for ourselves. I had not lived there a week before my feet wore a path from my door to the pond-side; and though it is five or six years since I trod it, it is still quite distinct. It is true, I fear that others may have fallen into it, and so helped to keep it open. The surface of the earth is soft and impressible by the feet of men; and so with the paths which the mind travels. How worn and dusty, then, must be the highways of the world, how deep the ruts of tradition and conformity! I did not wish to take a cabin passage, but rather to go before the mast and on the deck of the world, for there I could best see the moonlight amid the mountains. I do not wish to go below now.

Walden, or Life in the Woods

Contributor: Mike Stratton

If I Wished to See a Mountain
Henry David Thoreau

If I wished to see a mountain or other scenery under the most favorable auspices, I would go to it in the foulest of weather so as to be there when it cleared up. We are then in the most suitable mood and nature is most fresh and inspiring. There is no serenity so fair as that which is just established in a tearful eye.

On crossing Moosehead Lake in his birch bark canoe

Contributor: Mike Stratton

If You Judge Safety
Colin Fletcher

But if you judge safety to be the paramount consideration in life you should never, under any circumstances, go on long hikes alone. Don't take short hikes alone, either — or, for that matter, go anywhere alone. And avoid at all costs such foolhardy activities as driving, falling in love, or inhaling air that is almost certainly riddled with deadly germs. Wear wool next to the skin. Insure every good and chattel you possess against every conceivable contingency the future might bring, even if the premiums half-cripple the present. Never cross an intersection against a red light, even when you can see that all roads are clear for miles. And never, of course, explore the guts of an idea that seems as if it might threaten one of your more cherished beliefs. In your wisdom you will probably live to a ripe old age. But you may discover, just before you die, that you have been dead for a long, long time.

The Complete Walker

Contributor: Mike Stratton

If You Would Have a Lovely Garden
Author Unknown

If you would have a lovely garden, you will have a lovely life.

Shaker poem

Contributor: Jim Schoel

Joy
William Blake

He who binds to himself a joy
Does the winged life destroy.
But he who kisses the joy as it flies
Lives in eternity's sunrise.

Several Questions Answered

Contributor: Susan Schaefer

Let Your Love
Author Unknown

Let your love be like the misty rain, coming softly but flooding the river.

Contributor: Jim Schoel

Life Begins
Henry Miller

Life only begins when one drops below the surface, when one gives up the struggle, sinks and disappears from sight.

The Wisdom of the Heart

Contributor: Todd Tinkham

Little Things
Author Unknown

Most of us
miss out
on life's
big prizes.
The Pulitzer.
The Nobel.
Oscars.
Tonys.
Emmys.
But we're
all eligible
for life's
small pleasures.
A pat
on the back.
A kiss
behind the ear.
A four-pound bass.
A full moon.
An empty
parking space.
A crackling fire.
A great meal.
A glorious sunset.
Hot soup.
Cold beer.
Don't fret
about
copping life's
grand awards.
Enjoy its
tiny delights.
There are plenty
for all of us.

Contributor: Julie Brynteson

Magic in the Feel of a Paddle
Sigurd F. Olson

There is magic in the feel of a paddle and the movement
of a canoe, a magic compounded of distance, adventure,
solitude, and peace. The way of a canoe is the way of the
wilderness and of a freedom almost forgotten. It is an
antidote to insecurity, the open door to waterways of ages

past and a way of life with profound and abiding satisfactions. When a man is part of his canoe, he is part of all that canoes have ever known.

The Singing Wilderness

Contributor: Mike Stratton

Making a Difference
Anonymous

As the old man walked the beach at dawn, he noticed a young man ahead of him picking up starfish and flinging them into the sea.

Finally, catching up to the youth, he asked him why he was doing this. The answer was that the stranded starfish would die if left in the morning sun.

"But the beach goes on for miles and there are millions of starfish," countered the other. "How can your effort make any difference?" The young man looked at the starfish in his hand and then threw him safely in the waves. "It makes a difference to this one," he said.

Contributor: Charlie Harrington

Men and Women
Anne Morrow Lindbergh

Perhaps both men and women in America may hunger, in our material, outward, active, masculine culture, for the supposedly feminine qualities of heart, mind and spirit — qualities which are actually neither masculine nor feminine, but simply human qualities that have been neglected. It is growth along these lines that will make us whole, and will enable the individual to become a world to himself.

Gift from the Sea

Contributor: Jim Schoel

Men Pass Away
Sir Francis Drake

Men pass away but people abide. See that ye hold fast the heritage we leave you. Teach your children its value that, never in the coming centuries, their hearts may fail them or their hands grow weak. Hitherto we have been too much afraid — henceforth we will fear only God.

Contributor: Mike Stratton

Mountaineering
James Ramsey Ullman

There has been no time in human history when mountains and mountaineering have had so much to offer to men. We need to rediscover the vast, harmonious pattern of the natural world of which we are a part — the infinite complexity and variety of its components, the miraculous simplicity of the whole. We need to learn again those essential qualities in our own selves which make us what we are: the energy of our bodies, the alertness of our minds; curiosity and the desire to satisfy it, weakness and will to master it. The mountain way may well be a way of escape — from turmoil and doubt, from war and the threat of war, from the perplexities and sorrows of the artificial world we have built ourselves to live in. But in the truest and most profound sense, it is an escape not from but to reality.

Age of Mountaineering

Contributor: Mike Stratton

One Can't Take a Breath
Woodrow Wilson Sayre

One can't take a breath large enough to last a lifetime; one can't eat a meal big enough so that one never needs to eat again. Similarly, I don't think any climb can make you content never to climb again...Certainly, there are such values as warm friendship tested and strengthened through shared danger, the excitement of obstacles overcome by one's own efforts, or the beauty of the high, quiet places of the world. But these values can't be stored like canned goods. They need to be experienced live — many times.

Four Against Everest

Contributor: Mike Stratton

People Are Always Blaming
George Bernard Shaw

People are always blaming their circumstances for what they are. I don't believe in circumstances. The people who get on in this world are the people who get up and look for the circumstances they want, and if they can't find them, make them.

Mrs. Warren's Profession

Contributor: Mike Stratton

Placing Your Bets
Willi Unsoeld

Another question that has fascinated me is, "Where are you placing your bets?" Are you laying your chips on nature? Is the world of nature where it's at? Because it seems like it, that's where I've tried to take you. Or, are you laying your chips on Man? What's your ultimate value locus? Imagine yourself God. You're reared back to produce a miracle. You're going to create, but you've got two ideas in mind. You're going to create either an earth with water and plant life and animals, or you're going to produce an earth with water and soil and flowers and mountains and animals and men, you know; and you can choose either one you want. Earth without man or earth with man. Let's have a show of hands. How many would create earth without man? How many, earth with man? I've agonized over this, but I come out very clearly on the side of man. I'm man-centered when it comes to value; it's probably because I was born one. Therefore, it provides you with a final test. Why don't you stay in the wilderness? Because that isn't where it's at; it's back in the city, back in downtown St. Louis, back in Los Angeles. The final test is whether your experience of the sacred in nature enables you to cope more effectively with the problems of man. If it does not enable you to cope more effectively with the problems — and sometimes it doesn't, sometimes it just sucks you right out into the wilderness and you stay there the rest of your life — then when that happens, by my scale of value, it's failed. You go to nature for an experience of the sacred; and I point out to you that it is not the only place that one can go, but in the Outward Bound and in my own experience it's the one that tends to be emphasized. You go there to re-establish your contact with the core of things, where it's really at, in order to enable you to come back into the world of man and operate more effectively. So I finish with the principle: Seek ye first the kingdom of nature that the kingdom of man might be realized.

Contributor: Mike Stratton

Psalm 23
David

> The Lord is my shepherd, I shall not want;
> He makes me lie down in green pastures;
> He leads me beside still waters.
> He restores my soul;
> He leads me in the paths of
> righteousness for his name's sake.

Even though I walk through the valley
of the shadow of death, I will
fear no evil;
For thou art with me; thy rod and thy
staff, they comfort me.
Thou preparest a table before me
in the presence of my enemies;
Thou anointest my head with oil, my cup overflows.
Surely goodness and mercy shall
follow me all the days of my life;
And I shall dwell in the house of the Lord forever.

The Bible

Contributor: Mike Stratton

Pursuit of Excellence

Author Unknown

The pursuit of excellence is gratifying and healthy. The pursuit of perfection is frustrating and neurotic. It's also a terrible waste of time.

Contributor: Jim Schoel

Quality

Robert Pirsig

A person who knows how to fix motorcycles — with Quality — is less likely to run short of friends than one who doesn't. And they aren't going to see him as some kind of object either. Quality destroys objectivity every time.

Or if he takes whatever dull job he's stuck with — and they are all, sooner or later, dull — and, just to keep himself amused, starts to look for options of Quality, and secretly pursues these options, just for their own sake, thus making an art out of what he is doing, he's likely to discover that he becomes a much more interesting person and much less of an object to the people around him because his Quality decisions change him too. And not only the job and him, but others too, because the Quality tends to fan out like waves. The Quality job he didn't think anyone was going to see is seen, and the person who sees it feels a little better because of it and is likely to pass that feeling on to others, and in that way the Quality tends to keep on going. My personal feeling is that this is how any further improvements of the world will be done: by individuals making Quality decisions and that's all.

Zen and the Art of Motorcycle Maintenance

Contributor: Jim Schoel

Security of the Sea
Joseph Conrad

I came out again on the quarterdeck, agreeably at ease in my sleeping-suit on that warm breathless night, barefooted, a glowing cigar in my teeth, and, going forward, I was met by the profound silence of the fore end of the ship. Only as I passed the door of the forecastle I heard a deep, quiet, trustful sigh of some sleeper inside. And suddenly I rejoiced in the great security of the sea as compared with the unrest of the land, in my choice of that untempted life presenting no disquieting problems, invested with an elementary moral beauty by the absolute straightforwardness of its appeal and by the singleness of its purpose.

The Secret Sharer

Contributor: Susan Schaefer

The Seven Sins of the World
Mahatma Gandhi

Wealth without work.
Pleasure without conscience.
Knowledge without character.
Commerce without morality.
Science without humanity.
Worship without sacrifice.
Politics without principle.

The Reverend Howard Hunter

Social Diseases
Kurt Hahn

There can be no doubt that the young of today have to be protected against certain poisonous effects inherent in present-day civilization. Five social diseases surround them, even in early childhood. There is the decline in fitness, due to modern methods of locomotion; the decline in initiative, due to the widespread disease of spectatoritis; the decline in care and skill, due to the weakened tradition of craftsmanship; the decline in self-discipline, due to the ever-present availability of tranquilizers and the stimulants; the decline of compassion, which William Temple called "spiritual death."

Contributor: Jim Schoel

Sorrow Happens

Author Unknown

> Sorrow happens, hardship happens.
> The hell with it. Who never knew
> The price of happiness will not be happy.
> Forgive no error you recognize,
> It will repeat itself, increase,
> And afterwards our pupils
> Will not forgive in us what we forgave.

Contributor: Jim Schoel

Still Round the Corner

J.R.R. Tolkien

> Still round the corner there may wait
> A new road or a secret gate;
> And though I oft have passed them by,
> A day will come at last when I
> Shall take the hidden paths that run
> West of the Moon, East of the Sun.

The Return of the King

Contributor: Mike Stratton

Strats' Tower of Strength

Mike Stratton

COURAGE gives me the strength to put worthwhile ideas into action.

COMPETENCE is the ability to perform honestly the job for which I am suited.

CULTURE displays the belief that an appreciation of life's goodness is a source of joy forever.

COURTESY is the outward expression of inner respect for the individual.

CHARACTER is that spiritual force within me that demands and gets my best choices and my best efforts.

Contributor: Mike Stratton

The Great Plains

Ian Frazier

This, finally, is the punch line of our two hundred years on the Great Plains: we trap out the beaver, subtract the Mandan, infect the Blackfeet and the Hidatsa and the Assiniboin, overdose the Arikara; call the land a desert and hurry across it to get to California and Oregon; suck up the buffalo, bones and all; kill off nations of elk and wolves and cranes and prairie chickens and prairie dogs; dig up the gold and rebury it in vaults someplace else; ruin the Sioux and Cheyenne and Arapaho and Crow and Kiowa and Comanche; kill Crazy Horse, kill Sitting Bull; harvest wave after wave of immigrants' dreams and send the wised-up dreamers on their way; plow the topsoil until it blows to the ocean; ship out the wheat, ship out the cattle; dig up the earth itself and burn it in power plants and send the power down the line; dismiss the small farmers, empty the little towns; drill the oil and the natural gas and pipe it away; dry up the rivers and the springs, deep-drill for irrigation water as the aquifer retreats. And in return we condense unimaginable amounts of treasure into weapons buried beneath the land that so much treasure came from — weapons for which our best hope might be that we will someday take them apart and throw them away, and for which our next-best hope certainly is that they remain humming way under the prairie, absorbing fear and maintenance, unused, forever.

The New Yorker

Contributor: Jim Schoel

The Human Soul
Carl Jung

If one wishes to understand the human soul, he need not bother with experimental psychology of the laboratory, which can tell him practically nothing. He would be better advised to take off his academic robes, and wander with open heart through the world: through the horrors of prisons, insane asylums and hospitals, through dirty, dirty dives and houses of prostitution or gambling, through the drawing rooms of elegant society, the stock exchanges, the socialist meetings, the churches and revival meetings of the cults, to experience love and hate, passion in every form, in his own being...He will come back with wisdom which no five-foot shelf of textbooks could give him and he will be capable of being a doctor to the human soul.

Contributor: Mike Stratton

The Inner Life
Joel Goldsmith

It is not what our outer life is. We do not prepare for the acceptance of our spiritual identity merely by trying to change ourselves into good human beings. Being good humanly has no relationship whatsoever to "dying daily." "Dying daily" is a realization that we are dissatisfied with our present mode of life, dissatisfied even if we have economic sufficiency, good health, or a happy family. Still there is that dissatisfaction, that sense of something missing in us. There is an inner unrest, a lack of peace, an inner discontent.

Without this hunger and this inner drive, there is no "dying." But as soon as we make the decision that we are going to walk the way that leads to spiritual fulfillment, we have begun the necessary transformation of mind; we have begun our spiritual journey. First must come the clear-cut realization that we cannot go on being just human beings, and attempt to add God's grace to our humanhood. There must be a turning; there must be an inner transformation. This has nothing to do with our outer life. The change takes place within us. The whole experience is an inner experience; it is one of consciousness, but when it takes place, it affects our entire outer experience.

A Parenthesis in Eternity

Contributors: Jim Schoel and Barry Orms

The Non-Conformist
Todd Tinkham

In the conformist wilderness of our day, I will eagerly do whatever non-conforming activities exist, and do them well, not for any other reason than to set myself free from this stale-smelling shelter called civilization. So, if my daughter says, "Dad, let's climb a tree," I do not say, "I'm too old for that." I climb and, once up, realize that I was too young to have been so grounded. And if I ask, "Who will come with me to watch the sun rise?" And, one by one, my friends decline, saying, "I do not have the time," or "I have seen so many, they are all alike," then I will go alone to a quiet place and watch the sun burst out of the ocean and splatter the sky with colors of day. And watching, I will feel those same colors within me — and later use them to paint, with delicate strokes, the places I go and the people I meet.

Contributor: Todd Tinkham

The Pillars of Hahnism
Kurt Hahn

I regard it as the foremost task of education to insure the survival of these qualities:
- an enterprising curiosity
- an undefeatable spirit
- tenacity in pursuit

- readiness for sensible self-denial
- and, above all, compassion.

Contributor: Mike Stratton

Those Only are Happy
John Stuart Mill

Those only are happy who have their minds fixed on some subject other than their own happiness, on the happiness of others, on the improvement of mankind, even on some art or pursuit, followed not as a means but as itself an ideal end. Aiming thus at something else, they find happiness by the way.

Contributor: Ted Woodward

Until We Diminish Our Attachment
Peter Coyote

Until we diminish our attachment to progress, ease and newness, voluntarily assume more trials and labors of existence, we are whistling in the dark and fated to nurse puny spirits. Only by using the apparent advantages of machines sparingly and with ritual restraint will we regain a world that is once again scaled to human dimensions. We will replace machines with tight community and social forms that support, contain, protect, and necessarily pinch off indulgences that now pass for liberty. Using less, buying less, trusting imagination over glut; exploring boredom, anger, greed; learning to follow or lead as situations demand; understanding the distinctions between power and restraint — these are the avenues, I believe, to knowledge of verifiable worth. The one objective arbiter for this venture is nature, and all we know of Her are the life and laws of our bodies and this planet.

Co-Evolution Quarterly

Contributor: Jim Schoel

Walking
Paul Swatek

The most natural form of locomotion, walking, has been in use since before the invention of the wheel and discovery of fire. Reliable and totally non-polluting, it offers convenience, no parking, no cost. Invigorating, it promotes health and gives you the chance to think.

User's Guide to the Protection of the Environment

Contributor: Mike Stratton

We Are So Ruled
Leo Buscaglia

We are so ruled by what people tell us we must be that we have forgotten who we are.

Contributor: Karl Rohnke

We Do Not Believe in Ourselves
e.e. cummings

We do not believe in ourselves until someone reveals that deep inside us something is valuable, worth listening to, worthy of our trust, sacred to our touch. Once we believe in ourselves we can risk curiosity, wonder, spontaneous delight or any experience that reveals the human spirit.

Contributor: Jim Schoel

Why You Can't Stay
Willi Unsoeld

And so, what is the final test of the efficacy of this wilderness experience we've just been through together? Because having been there, in the mountains, alone, in the midst of solitude and this feeling, this mystical feeling, if you will, of the ultimacy of joy and whatever there is, the question is, "Why not stay out there in the wilderness the rest of your days and live in the lap of Satori or whatever you want to call it?"

And the answer, my answer, to that is, "Because that's not where men are." And the final test for me of the legitimacy of the experience is, how well does your experience of the sacred in nature enable you to cope more effectively with the problems of mankind when you come back to the city?

And now you see how this phases with the role of the wilderness. It's a renewal exercise and as I visualize it, it leads to a process of alternation. You go to nature for your metaphysical fix — your reassurance that the World makes sense. It's reassurance that I don't get in the city, but with that excess of confidence of reassurance that there is something behind it all and it is good. You come back to where men are, to where men are messing things up, because men tend to do that and you come back with a new ability to relate to yourself and to your fellow man and help your fellow men to relate to each other.

Contributor: Paul Radcliffe

Wind, Sand and Stars
Antoine de Saint-Exupéry

Old bureaucrat, my comrade, it is not you who are to blame. No one ever helped you to escape. You, like a termite, built your peace by blocking up with cement every chink and cranny through which the light might pierce. You rolled yourself up into a ball in your genteel security, in routine, in the stifling conventions of provincial life, raising a modest rampart against the winds and the tides and the stars. You have chosen not to be perturbed by great problems, having trouble enough to forget your own fate as man. You are not the dweller upon an errant planet and do not ask yourself questions to which there are no answers. You are a petty bourgeois of Toulouse. Nobody grasped you by the shoulder while there was still time. Now the clay of which you were shaped has dried and hardened, and naught in you will ever awaken the sleeping musician, the poet, the astronomer that possibly inhabited you in the beginning.

Wind, Sand and Stars

Contributor: Jim Schoel

You Had To Be Taught
Nicholas Johnson

You had to be taught to demand heavy doses of refined sugar in your diet, and other non-food products and stimulants: soft drinks, coffee, and alcohol. You had to be sold the idea of filling your lungs with smoke. You had to have your curiosity and creativity pounded out of you by your parents and teachers. You have to be conned into buying the junk that clutters up your life — decreasing your happiness and your pocketbook. You have to be trained to be physically lazy.

You're all right. You can live on the earth just fine. You were made for it. You're welcome there. It's all that stuff that's been laid on you — at some corporation's profit — that's making you miserable and causing you to lose track of the rhythm of the earth. You can get it back. But only when you remember where you lost it.

Test Pattern for Living

Contributor: Mike Stratton

You Ride a Bicycle
Nicholas Johnson

You ride a bicycle because it feels good. The air feels good on your body; even the rain feels good. The blood starts moving around your body, and pretty soon it gets to your head, and, glory be, your head feels good. You start noticing things. You look until you really see. You hear things, and smell smells you never knew were there. You start whistling nice little original tunes to suit the moment. Words start getting caught in the web of poetry in your mind. And there's a nice feeling too, in knowing you're doing a funda-mental life thing for yourself: transportation. You got a little bit of your life back! And the thing you use is simple, functional and relatively cheap. You want one that fits you and rides smoothly, but with proper care and few parts it should last almost forever. Your satisfaction comes from within you, not from the envy or jealousy of others (although you are entitled to feel a little smug during rush hours, knowing you are also making better time than most of the people in cars).

Test Pattern for Living

Contributor: Mike Stratton

Zen Story
Joe Hymas

Once a Zen master received a university professor who came to inquire about Zen. It was obvious to the master from the start of the conversation that the professor was not so much interested in learning about Zen as he was in impress-ing the master with his own opinions and knowledge. The master listened patiently and finally suggested they have tea. The master poured his visitor's cup full and then kept on pouring. The professor watched the cup overflowing until he could no longer restrain himself. "The cup is overfull, no more will go in." "Like this cup," the master said, "you are full of your own opinions and speculations. How can I show you Zen unless you first empty your cup?"

Zen in the Martial Arts

Contributor: Tim Churchard

ONE-LINERS

Growth

A Great Fighter
Sugar Ray Robinson

The mark of a great fighter is how he acts when he's getting licked.

Contributor: Tim Churchard

Conquering Self
Plato

The first and best victory is to conquer self; to be conquered by self is, of all things, the most shameful and vile.

Contributor: Jim Schoel

The Demands of Life
Dag Hammarskjöld

Life only demands from you the strength you possess. Only one feat is possible... not to have run away.

Markings

Contributor: Ted Woodward

Disability
Kurt Hahn

Your disability is your opportunity.

Contributor: Mike Stratton

Each Morning
Edna Hong

Each morning the child cups his hands and receives life, thumb-rim full, and lets fear slip through his little fingers.

Contributor: Susan Schaefer

The Educated Man
Sidney Harris

The truly educated man is the one who knows and can properly appraise the consequences of his actions.

Reality Therapy

Contributor: Jim Schoel

Educate a Human Being
Author Unknown

Those who want to leave an impression for one year should plant corn; those who want to leave an impression for ten years should plant a tree; but those who want to leave an impression for 100 years should educate a human being.

Ancient Chinese proverb

Contributor: Scott Garman

Experience
Oscar Wilde

Experience is the name everyone gives to their mistakes.

Contributor: Ted Woodward

Fire the Soul
La Fontaine

Man is so made, that whenever anything fires his soul... impossibilities vanish.

Contributor: Eric Schusser

For Experiences
Kurt Hahn

It is the sin of the soul to force young people into opinions — indoctrination is of the devil — but it is culpable neglect not to impel young people into experiences.

Contributor: Mike Stratton

It is the Weak Who Are Cruel
Leo Rosten

It is the weak who are cruel. Gentleness can only be expected from the strong.

Contributor: Mike Stratton

Many Arrivals
Theodore Roethke

Many arrivals make us live.

Contributor: Jim Schoel

Many Departures
Jim Schoel

You've got to leave before you can arrive.

Contributor: Jim Schoel

Never Be Ashamed
Alexander Pope

A man should never be ashamed to say he has been wrong, which is but saying in other words that he is wiser today than he was yesterday.

Contributor: Mike Stratton

Place to Improve the World
Robert M. Pirsig

The place to improve the world is first in one's own heart and head and hands, and then work outward from there.

Zen and the Art of Motorcycle Maintenance

Contributor: Jim Schoel

Play for Growth
Ashley Montagu

In the early formative years, play is almost synonymous with life. It is second only to being nourished, protected and loved. It is a basic ingredient of physical, intellectual, social and emotional growth.

Contributor: Jim Schoel

Saintliness is Out of Reach
Kurt Hahn

Saintliness is out of reach for most adults, even for bishops, so I do not think we should expect youngsters to be saints — capable of living up to their ideals merely through the strength of their conscience.

Contributor: Mike Stratton

Summer Within
Albert Camus

In the depth of winter, I finally learned that within me there lay an invincible summer.

Summer, 1954

Contributors: Lee and Robert Natti

Three Hardest Tasks
Sidney Harris

The three hardest tasks in the world are neither physical feats nor intellectual achievements, but moral acts: to return love for hate, to include the excluded, and to say, "I was wrong."

Contributor: Susan Schaefer

To Know Good
Sanskrit

Not everything is good because it is old nor poems always bad by being new. Good people try both before they make their choice, while the fool but takes the view of others.

Contributor: Mike Stratton

Want Little
Surya Prem

Teach children to want little while they are little.

Contributor: Todd Tinkham

The Way
Lao Tzu

The way to do is to be.

Contributor: Jim Schoel

We Have Suffered
Sir Ernest H. Shackleton, at South Georgia Island, 1916, while leading his men to safety after their ship was crushed in the ice in the Antarctic.

We have suffered, starved, and triumphed; grovelled down yet grasped at glory...grown bigger in the bigness of the whole.

Contributor: Mike Stratton

Outlook

Be Nice
Strats

It doesn't cost anything to be nice.

Contributor: Mike Stratton

Beer Cans Are Beautiful
Edward Abbey

Beer cans are beautiful — it's the highway that's ugly.

In regard to littering roadsides with beer cans

Contributor: Jim Schoel

Before Breakfast
Lewis Carroll

Sometimes I've believed as many as six impossible things before breakfast.

Contributor: Jim Schoel

Being Critical
Bryson Coles

Sometimes it's easier to be critical than to be right.

Contributor: Jim Schoel

Do Good
Eldress Harriet Bullard

Do good. This should be the aim of every human being, to make the world better for their having lived.

Contributor: Jim Schoel

Feelings
Mike Donovan

Feelings aren't a life sentence.

ABC workshop participant

Contributor: Mike Donovan

Happiness
Abraham Lincoln

Most folks are about as happy as they make up their minds to be.

Contributor: Mike Stratton

He Who Has a Why
Frederick Nietzsche

He who has a why to live can bear with almost any how.

Contributor: Jim Schoel

Imagination
Albert Einstein

Imagination is more important than knowledge.

Contributor: Luke Schoel

Let Me Live
Sir Rabindranath Tagore

Only let me make my life simple and straight like a flute of reed for thee to fill me with music.

Contributor: Mike Stratton

Life is Serious
Ancient Chinese Philosopher

Life is serious, but not that serious.

Contributor: Bonnie Hannable

Life is Tragic
Author Unknown

Life is tragic to those who feel and comic to those who think.

Contributor: Jim Schoel

Madness
Seneca

There is no genius free from tincture of madness.

Contributor: Todd Tinkham

Murphy's Law
Peter Willauer

Expect that which you dread — the inevitability of the unwanted!

Contributor: Peter Willauer

Only the Best?
Henry David Thoreau

The world would be very silent if no birds sang there except those that sang the best.

Contributor: Jim Schoel

Point of View
Author Unknown

The pessimist looks at opportunities and sees difficulties. The optimist looks at difficulties and sees opportunities.

Contributor: Jim Schoel

Roots
Marcus Tullius Cicero

In all great arts as in trees it is the height that charms us; we care nothing for the roots, or trunks, yet it could not be without the aid of these.

Contributor: Scott Garman

Ropes Courses
Mike Stratton

The ropes course is the Swiss army knife of Outdoor Education.

Contributor: Karl Rohnke

Rose Bushes
Abraham Lincoln

We can complain because rose bushes have thorns, or rejoice because thorn bushes have roses.

Contributor: Mike Stratton

Thunder and Lightning
Mark Twain

Thunder is good. Thunder is impressive, but it is lightning that does the work.

Contributor: Ted Woodward

To Ponder
John Charles Amesse

Some are chosen to ponder while the world flits by without a glance.

Contributor: Mike Stratton

Relationships

A Friend
Anonymous

A friend is a present that you give yourself.

Contributor: Ted Woodward

Being Alone
Ellen Burstyn

What a lovely surprise to finally discover how unlonely being alone can be.

Contributor: Scott Garman

The Body
Author Unknown

Every man is a builder of a temple called his body. Keep your body high on the list of environments to protect from pollution.

Contributor: Mike Stratton

Bond of the Sea
Joseph Conrad

Between the five of us was the strong bond of the sea, and also the fellowship of the craft, which no amount of enthusiasm for yachting, cruising and so on can give, since one is only the amusement of life and the other is life itself.

Youth

Contributor: Peter Coburn

Everything in the Universe
John Muir

When we try to pick out anything by itself we find it hitches to everything in the universe.

Contributor: Jim Schoel

Helping Others
Elie Wiesel

Does not helping another person mean saving him from despair?

Contributor: Jim Schoel

How to Serve
Albert Schweitzer

One thing I know: the only ones among you who will be really happy are those who will have sought and found how to serve.

Contributor: Jim Schoel

Learn to be Better
Wendell Berry

It is not from ourselves that we will learn to be better than we are.

A Native Hill

Contributor: Todd Tinkham

Minority
Herbert Prochnow

The minority is always wrong, at the beginning.

Contributor: Todd Tinkham

Self-Estimation
Henry David Thoreau

Public opinion is a weak tyrant compared with our own private opinion. What a man thinks of himself, that it is which determines, or rather indicates, his fate.

Contributor: Mike Stratton

Solitude
Søren Kierkegaard

All evil stems from this: that men do not know how to handle solitude.

Contributor: Susan Schaefer

Walls Instead of Bridges
Author Unknown

People are lonely because they build walls instead of bridges.

Contributor: Jim Schoel

Who Owns the Earth
Anonymous

We do not inherit the earth from our ancestors, we borrow it from our children.

Contributor: Mike Stratton

Taking Risks

Bravest Man
Kurt Hahn

The bravest man is the one who weighs up all the risks and when they have become greater than the object is worth, has the courage to turn back and to face the other risk of being called a coward.

Contributor: Mike Stratton

Consequences
R.G.Ingersoll

In nature there are neither rewards nor punishments. There are consequences.

Contributor: Ted Woodward

Dare Mighty Things
Theodore Roosevelt

Far better it is to dare mighty things, to win glorious triumphs, even though checkered by failure, than to rank with those poor spirits who neither enjoy much nor suffer much, because they live in the gray twilight that knows no victory nor defeat.

While Daring Greatly

Contributor: Jim Schoel

Decision
Eric Langmuir

A decision without the pressure of consequence is hardly a decision at all.

Contributor: Ted Woodward

Doing the Impossible
Walter Bagehof

The great pleasure in life is doing what people say you cannot do.

Contributor: Mike Stratton

Fame
Reggie Jackson

Fans don't boo nobodies!

Contributor: Mike Stratton

Fear Not Death
Marcus Aurelius

It is not death that a man should fear, but he should fear never beginning to live.

Contributor: Ted Woodward

It Can Be Fun
Mike Stratton

It can be fun to...bust your ass.

Contributor: Mike Stratton

It is Required of a Man
Oliver Wendell Holmes

I think, that as life is action and passion, it is required of a man that he should share the passion and action of his time, at peril of being judged not to have lived.

Contributor: Mike Stratton

It Takes Courage
Paul Tillich

Courage is a greater virtue than love. At best, it takes courage to love.

Contributor: Mike Stratton

The Last Bridge
Heinrich Harrer

I believe that no man can be completely able to summon all his strength, all his will, all his energy, for that last desperate move, till he is convinced the last bridge is down behind him and that there is nowhere to go but on.

Contributor: Jim Schoel

Letting Go
Anonymous

Courage is the power to let go of the familiar.

Contributor: Jim Schoel

Look...Don't Hesitate
Author Unknown

You should look before you leap, but he who hesitates is lost.

Contributor: Karen Lofthus Carrington

Moral Courage
Mark Twain

It is curious that physical courage should be so common in the world, and moral courage so rare.

Contributor: Mike Stratton

Play
Holbrook Jackson

We take chances, risk great odds, love, laugh, dance...in short, we play. The people who play are the creators.

Contributor: Mike Stratton

Play the Game
Winston Churchill

Play the game for more than you can afford to lose...only then will you learn the game.

Contributor: Mike Stratton

Serious Daring
Eudora Welty

All serious daring starts from within.

One Writer's Beginnings

Contributor: Scott Garman

Stand for Something
Anonymous

If you don't stand for something, you'll fall for anything.

Contributor: Mike Stratton

To Venture
Søren Kierkegaard

To venture causes anxiety, but not to venture is to lose one's self. And to venture in the highest sense is precisely to become conscious of one's self.

Contributor: Jim Schoel

The Triumph of Evil
Edmund Burke

All that is necessary for the triumph of evil is that good men do nothing.

Thoughts on the Cause of the Present Discontents

April 23, 1770

Contributor: Mike Stratton

Warning Lights
B. Von Alten

Environmentalists are the warning lights on the dashboard of civilization.

Contributor: Al Katz

What's Important
Barry Goldwater

I have just about reached the conclusion that, while large industry is important, fresh air and clean water are more important, and the day may well come when we have to lay that kind of hand on the table and see who is bluffing.

Contributor: Jim Schoel

SONGS

Amazing Grace
John Newton

> Amazing grace, how sweet the sound,
> That saved a wretch like me.
> I once was lost, but now am found,
> Was blind, but now I see.
>
> 'Twas grace that taught my heart to fear,
> And grace my fears relieved,
> How precious did that grace appear
> The hour I first believed.
>
> Through many troubles, toils and snares
> I have already come.
> 'Tis grace hath brought me safe thus far
> And grace will lead me home.
>
> When we've been there ten thousand years
> Bright shining as the sun,
> We've no less days to sing God's praise
> Than when we'd first begun.

Contributor: Jim Schoel

For the Beauty of the Earth
F.S. Pierpont

> For the beauty of the earth,
> For the glory of the skies,
> For the love which from our birth,
> Over and around us lies, Refrain:
>
> *Refrain:*
> Lord of all, to thee we raise
> This our hymn of grateful praise.

For the beauty of each hour
Of the day and of the night,
Hill and vale, and tree and flower,
Sun and moon, and stars of light, (Refrain)

For the joy of ear and eye,
For the heart and mind's delight,
For the mystic harmony
Linking sense to sound and sight, (Refrain)

For the joy of human love,
Brother, sister, parent, child,
Friends on earth, and friends above,
For all gentle thoughts and mild, (Refrain)

Contributor: Mike Stratton

He's Got the Whole World
Traditional

He's got the whole world in his hands,
He's got the whole wide world in his hands,
He's got the whole world in his hands,
He's got the whole world in his hands.

He's got the tiny little baby in his hands,
(Repeat twice) He's got the whole world in his hands.

He's got you and me, brother, in his hands,
He's got you and me, sister, in his hands,
He's got everybody here in his hands,
He's got the whole world in his hands.

Contributor: Jim Schoel

Home on the Range
Traditional

Oh! Give me a home where the buffalo roam,
Where the deer and the antelope play;
Where seldom is heard a discouraging word,
And the skies are not cloudy all day.

Chorus:
Home, home on the range!
Where the deer and the antelope play,
Where seldom is heard a discouraging word,
And the skies are not cloudy all day.

Oh! Give me a land where the bright diamond sand
Throws its light from the glittering streams;
Where glideth along the graceful white swan,
Like the maid to her heavenly dreams. (Chorus)

How often at night, when the heavens were bright,
With the light of the twinkling stars,
Have I stood here amazed, and asked as I gazed
If their glory exceeds that of ours. (Chorus)

I love the wild flowers in this bright land of ours,
And I love the wild curlew's shrill scream;
The bluffs and white rocks and antelope flocks,
That graze on the mountain so green. (Chorus)

Contributor: Bud Patterson

Simple Gifts
Author Unknown

'Tis the gift to be simple, 'tis the gift to be free,
'Tis the gift to come down where we ought to be,
And when we find ourselves in the place just right,
'Twill be in the valley of love and delight.

When true simplicity is gain'd,
To bow and to bend we shan't be ashamed,
To turn, turn will be our delight
Till by turning, turning we come round right.

Shaker tune

Contributor: Jim Schoel

Swing Low, Sweet Chariot
Traditional

Chorus
Swing low, sweet chariot,
Comin' for to carry me home!
Swing low, sweet chariot,
Comin' for to carry me home!

I looked over Jordan,
An' what did I see,
Comin' for to carry me home!
A band of angels comin' after me,
Comin' for to carry me home! (Chorus)

If you get to heaven before I do,
Comin' for to carry me home!
Just tell my friends
I'm comin' too,
Comin' for to carry me home! (Chorus)

I'm sometimes up and sometimes down,
Comin' for to carry me home!
But still my soul feels heavenly bound,
Comin' for to carry me home! (Chorus)

Contributor: Mike Stratton

You Are My Sunshine
Traditional

You are my sunshine, my only sunshine.
You make me happy when skies are gray.
You'll never know, dear,
How much I love you.
Please don't take my sunshine away.

The other night, dear, as I lay sleeping
I dreamed I held you in my arms.
When I awoke, dear,
I was mistaken
And I hung my head and cried.

I'll always love you and make you happy
If you will only say the same,
But if you leave to
Love another
You'll regret it all some day.

You told me once, dear, you really loved me
And no one else could come between.
But now you've left me,
And love another.
You have shattered all my dreams.

Contributor: Jim Schoel

CONTRIBUTORS

Celeste Archambault, former HIOBS instructor, founding teacher in Gloucester Museum School, now resides in Seattle, Washington

George Ashley, Phillis Wheatley Association Ropes Program, Cleveland, Ohio

Beau Bassett, former Outward Bound Instructor, Director of the American Youth Foundation's Camp Merrowvista, Ossipee, New Hampshire

Dan Baker, teacher at Rochester Alternative School, Rochester, New Hampshire

Carl Brown, former HIOBS instructor, ship's captain, playwright.

Julie Brynteson, former Bounders instructor, Lincoln, Massachusetts

Karen Lofthus Carrington, Jungian Therapist, Woodacre, California

Tim Churchard, co-founder of Rochester Alternative School, instructor at the University of New Hampshire, W. Lebanon, Maine

Peter Coburn, President of Commercial Logic Software Company, Outward Bound instructor at HIOBS and North Carolina, Norwich, Vermont

Arthur Conquest III, Educational Consultant and former Outward Bound Instructor at HIOBS and other schools, Brookline, Massachusetts

Bill Cuff, PA Trainer, co-founder of Rochester Alternative School, Doctoral student, Milton, New Hampshire

Mike Donovan, teacher, Glen Grove School, Glenview, Illinois

Rufus Little, former PA Staff, Director of Athletics, Choate School, Wallingford, Connecticutt

Bill Frankel, Sargent Camp (Boston University), Peterborough, New Hampshire

Howie Futterman, Adaptive Physical Education Teacher, New York City Public Schools, Flushing, Queens

Lisa Galm, PA Staff, Director of PA Challenge Program, Covington, Georgia

Scott Garman, former PA Staff, PA Trainer, Japan

Mike Gass, professor at University of New Hampshire, board member of Association of Experiential Education, Durham, New Hampshire

George Gorman, Therapist, Minister, Gloucester, Massachusetts

Bonnie Hannable, PA Staff, Administrative Assistant, Hamilton, Massachusetts

Charlie Harrington, PA Trainer, teacher in the Arlington Public Schools, Arlington, Massachusetts

Karen Hegeman, teacher, Newark High School, Newark, New York

Bob Henderson, Faculty of Education, Queen's University at Kingston, Ontario, Canada

Laurel Hood, PA Intern, 1989, Kingston, Ontario, Canada

Bert Horwood, Faculty of Education, Queen's University at Kingston, Ontario, Canada

Kathy Hunt, PA Trainer, former PA Staff, Orono, Maine

Wendy Hutchinson, Frederick Douglass Middle School, Rochester, New York

Sally Jepson, teacher, Portland, Oregon

Al Katz, former PA Staff, outdoor educator and ropes course builder, Warren, New Jersey

Bruce Kezlarian, Groves High School, Birmingham, Michigan

Will Lange III, former Director of Dartmouth Outward Bound School.

Lance Lee, former Outward Bound Instructor, founding Director of Rockport Apprenticeshop, Rockport, Maine

Don MacDowall, Camping Association of Victoria, Australia

Dr. Richard Maizell, Program Coordinator of Project Quest, Sussex County Vocational School, PA Trainer, Hopatcong, New Jersey

Joanne Maynard, PA Trainer, Special Needs Teacher, Marblehead Public Schools, Gloucester, Massachusetts

Jim Moll, Director of the Experiential Education Center, Wylie E. Groves High School, Birmingham, Michigan

Mark Murray, PA Staff, Hamilton, Mass.

Lee and Robert Natti, writer and school principal respectively, Gloucester, Massachusetts

Rich Obenschain, Director of the Gordon College La Vida program, Wenham, Massachusetts

Barry Orms, Management Consultant, Dick Scott Entertainment, and Boys Harbor, New York, New York

Bud Patterson, Alternative School Teacher, Cheyenne, Wyoming

Rafe Parker, former HIOBS course director, former Director of the SWOBS, current Director of Sea Education Associates, Woods Hole, Massachusetts

Brian Pritchard, Vermont Office on Aging, Waterbury, Vermont

Paul Radcliffe, PA Staff, Hamilton, Massachusetts

James Raffan, Queen's University Faculty of Education, Ontario

Bob Rheault, Director of Vietnam Veterans Program, HIOBS, Rockland, Maine

Nicole Richon-Schoel, Community Outreach coordinator, Wellspring House, Gloucester, Massachusetts

Karl Rohnke, President, PA Inc., Hamilton, Massachusetts

Bob Ryan, PA Staff, Hamilton, Massachusetts

Susan St. John-Rheault, HIOBS Instructor and Course Director, Rockland, Maine

Phil Salzman, teacher and coordinator of Health Education, Gloucester Public Schools, Gloucester, Massachusetts

Jay Sanderford, Minister, First Presbyterian Church, Charlottesville, Virginia

Susan Schaefer, PA Staff, Hamilton, Massachusetts

Eric Schusser, Dunstan High School, Alexandra, New Zealand

Luke Schoel, student and adventurer, Gloucester, Massachusetts

Pat Sheckler, Graham & Parks Alternative School, Somerville, Massachusetts

Sarah Smeltzer, Coordinator of Pine Lake Environmental Center, Hartwick College, Oneonta, New York

Ann Smolowe, PA Staff, Hamilton, Massachusetts

John K. Spencer, former instructor at HIOBS, computer consultant, Rockport, Massachusetts

Jemique Straker, Youth Leader, Boys Harbor, New York, New York

Nancy Stratton, former teacher at the Carroll School, Lincoln, Massachusetts

Todd Tinkham, PA practitioner, teacher, poet, Gloucester, Massachusetts

William Toomey, Director of Enterprise Co-op, an alternative program in the Cambridge, Massachusetts Schools.

Shelia Torbert, PA Staff, Covington, Georgia

Peter Willauer, founding Director of the Hurricane Island Outward Bound School, Rockland, Maine

Conrad Willeman, PA Staff, Hamilton, Massachusetts

Ted Woodward, former Outward Bound Instructor, teacher of adventure at Andrus Children's Home, Yonkers, New York.

Tom Zierk, PA Staff, Hamilton, Massachusetts

ACKNOWLEDGEMENTS

Speech to environmentalists in Missoula, Montana, by Edward Abbey, 1978, and other quotes reprinted by permission of Don Congdon Associates, Inc.

Address at Queen's University, by Donald Schön, Professor, Department of Urban Planning at the Massachusetts Institute of Technology. Reprinted by permission. Comments reprinted by permission of James Raffan.

A Light in the Attic, by Shel Silverstein. Copyright 1981 by Evil Eye Music, Inc. Reprinted by permission of Harper & Row, Publishers, Inc.

All I Ever Really Needed to Know I Learned in Kindergarten, as it appeared in THE KANSAS CITY TIMES. Copyright © 1986 by Robert Fulghum. Reprinted from *All I Ever Really Needed to Know I Learned in Kindergarten*, by Robert Fulghum, by permission of Villard Books, a Division of Random House, Inc.

Age of Mountaineering, by James Ramsey Ullman. © 1941, 1954 by J.R. Ullman.

Another Land Made of Water, by Gordon Bok. Reprinted by permission of Folk Legacy Records. Comments reprinted by permission of Peter Coburn.

Being First, a poem by Dan Baker. Used by permission.

Beware of Me, My Friend. Shelia Torbert's comments used by permission.

Circles on the Water, by Marge Piercy. Copyright © 1982 by Marge Piercy. Reprinted by permission of Alfred A. Knopf Inc.

The Collected Poems of Theodore Roethke, by Theodore Roethke. *Meditation at Oyster River, part 4.* Copyright © 1960 by Beatrice Roethke, Administratrix of the Estate of Theodore Roethke. *The Sloth*, Copyright © 1950 by Theodore Roethke. Used by permission of Doubleday, a division of Bantam, Doubleday Dell publishing Group, Inc. Comments used by permission of Joanne Maynard.

Commitment, by W.H. Murray. Comments used by permission of Rafe Parker.

The Complete Walker: The Joys and Techniques of Hiking and Backpacking, by Colin Fletcher. Copyright © 1968 by Alfred A. Knopf, Inc. Reprinted by permission.

Dandelion Wine, by Ray Bradbury. Copyright © 1956.

Deo Gratias, a poem by John A. Galm. Reprinted by permission of the author.

The Different Drum: Community Making and Peace, by M. Scott Peck, M.D. Copyright © 1988. Reprinted by permission of Touchstone Books.

Dirt, a poem by Robert Service. Comments used by permission of Joanne Maynard.

Disposable Society, lyrics by Eugene McDaniels. Reprinted by permission of Skyforest Music.

Enviropimp, by Randy Bayliss. From a letter to a friend, 1989. Reprinted by permission.

The Fire Next Time, by James Baldwin, copyright © 1962, 1963 by James Baldwin. Used by permission of Doubleday, a division of Bantam, Doubleday Dell Publishing Group, Inc.

Four Against Everest, by Woodrow Wilson Sayre. Copyright © 1964. Reprinted by permission of the publisher, Prentice-Hall, Inc., Englewood Cliffs, New Jersey. Comments used by permission of Phil Salzman.

Gift from the Sea, by Anne Morrow Lindbergh. Pantheon Books, a division of Random House. Reprinted by permission.

Go For It!, by Sharon Baack, 1982. Reprinted by permission of the author and the Baptist Sunday School Board, Nashville, TN.

Go For the Perfect Try, poem and comments by Joe Petriccione and Sarah Smeltzer. Used by permission.

The Great Plains, by Ian Frazier. Copyright © 1989 by Ian Frazier. Reprinted by permission of Farrar, Straus & Giroux, Inc.

Hahn, Kurt. All quotes reprinted by permission of Outward Bound, U.S.A.

Heads and Lives, by Phil Salzman. Reprinted by permission.

Holy Sweat, by Tim Hansel. Copyright © 1987 by Tim Hansel. Reprinted by permission of Word Books, Dallas, Texas.

The Hundredth Monkey, by Ken Kesey, as printed in the NEW PERSPECTIVE JOURNAL.

HIOBS Instructor's Manual, by Peter Osborne Willauer, Founder, HIOBS. Reprinted by permission.

Irrational Man; A Study in Existential Philosophy, by William Barrett, ©1958. Reprinted by permission of Doubleday, a division of Bantam, Doubleday Dell Publishing Group, Inc.

I Never Held Your Hand, by Tim Churchard. Used by permission.

If I Had My Life To Live Over, from the AHP PERSPECTIVE, 1975. Reprinted by permission of the ASSOCIATION FOR HUMANISTIC PSYCHOLOGY PERSPECTIVE, San Francisco.

Ithaka, by C.P. Cavafy. Comments used by permission of John K. Spencer.

Joys and Sorrows: reflections by Pablo Casals, as told to Albert E. Kahn. Copyright © 1970. Reprinted by permission of Simon & Schuster, Inc.

Lake Wobegon Days, by Garrison Keillor. Copyright © Garrison Keillor, 1985. Reprinted by permission.

The Learning Gate, by Carol Cornwell. Reprinted by permission. Comments used by permission of Bill Toomey.

Leaves of Grass, by Walt Whitman. Comments used by permission of Susan St. John-Rheault.

Letter to Danny, by Bill Cuff. Used by permission.

Life Together, by Dietrich Bonhoeffer. Copyright © 1954 by Harper & Row. Reprinted by permission of Harper & Row Publishers, Inc. Comments used by permission of Susan Schaefer.

The Little Prince, by Antoine de Saint-Exupéry. Copyright © 1943 by Antoine de Saint-Exupery and renewed 1971 by Harcourt Brace Jovanovich, Inc., reprinted by permission of the publisher and William Heinemann Ltd., Publishers, London.

Look For a Reaction, by Jemique Straker. Used by permission.

Man's Search for Meaning. Excerpt from the Preface by Gordon W. Allport. Reprinted by permission of Beacon Press.

The Tao of Leadership: Lao Tzu's Tao Te Ching Adapted for a New Age, by John Heider. Reprinted by permission of Humanics Limited.

Teacher & Child, by Haim Ginott. Copyright © 1972 by Macmillan Co. Reprinted by permission of Dr. Alice Ginott, executrix of the Estate of Haim Ginott.

Thank You for the Nourishment, by Paul Radcliffe. Used by permission.

To Build a Wood Boat, by Jim Schoel. From *Hanging in There* (an unpublished book of poems). Used by permission.

Ulysses, by Alfred, Lord Tennyson. Comments used by permission of Bob Rheault.

Until We Diminish Our Attachment, by Peter Coyote. Excerpt from COEVOLUTION QUARTERLY, Fall 1978. Reprinted by permission of THE WHOLE EARTH REVIEW and the author.

W.S. Coffin's Remarks on Service. Excerpt from the HIOBS Newsletter, Fall 1988. Reprinted by permission.

What Are The Qualities Of Life?, by Mike Gass, June 1982. Reprinted by permission.

Where the Sidewalk Ends, by Shel Silverstein. Copyright 1974 by Evil Eye Music, Inc. Reprinted by permission of Harper & Row, Publishers, Inc.

Wind, Sand and Stars, copyright © 1939 by Antoine de Saint-Exupéry and renewed by Lewis Galantiere. Reprinted by permission of Harcourt Brace Jovanovich, Inc., and William Heinemann Ltd., Publishers, London.

The Wisdom of the Heart, by Henry Miller. Copyright 1959 by Henry Miller. Copyright © 1959 by New Directions Publishing Company. Reprinted by permission.

You Can't Go Home Again, by Thomas Wolfe. Copyright © 1934, 1937, 1938, 1939, 1940 by Maxwell Perkins as the Executor of the Estate of Thomas Wolfe. Reprinted by permission of Harper & Row, Publishers, Inc.

Zen and the Art of Motorcycle Maintenance, by Robert Pirsig. Copyright © 1974 by the author. Reprinted by permission of William Morrow & Co.

Zorba the Greek, by Nikos Kazantzakis. Copyright © 1953, 1981 by Simon & Schuster, Inc. Reprinted by permission of Simon & Schuster, Inc.

INDEX A: NUGGET AUTHORS

L

M

N

O

P

R

S

T

U

V

W

Y

Z

Index B: Nugget Titles

C

D

E

I

J

K

L

M

T

U

W

Y

Z